Molecular Biology Intelligence Unit

Tyrosine Kinases and Neoplastic Transformation

Stuart Kellie, Ph.D.

Yamanouchi Research Institute
Littlemore Hospital
Oxford, United Kingdom

R.G. Landes Company
Austin

MOLECULAR BIOLOGY INTELLIGENCE UNIT
TYROSINE KINASES AND NEOPLASTIC TRANSFORMATION
R.G. LANDES COMPANY
Austin

CRC Press is the exclusive worldwide distributor of publications
of the Molecular Biology Intelligence Unit.
CRC Press, 2000 Corporate Blvd., NW, Boca Raton, FL 33431. Phone: 407/994-0555.

Submitted: August 1994
Published: November 1994

Please address all inquiries to the Publisher:
R.G. Landes Company, 909 Pine Street, Georgetown, TX 78626
or
P.O. Box 4858, Austin, TX 78765

Phone: 512/ 863 7762; FAX: 512/ 863 0081

ISBN 1-57059-075-3
CATALOG # LN9075

Library of Congress Cataloging-in-Publication Data

Kellie, Stuart
Tyrosine kinases and neoplastic transformation / Stuart Kellie.
p. m.—(Molecular biology intelligence unit)
Includes bibliographical references and index.
ISBN 1-57059-075-3 (hard cover)
1. Carcinogenesis. 2. Protein-tyrosine kinase. 3. Cell transformation. 4. Oncogenes.
I. Title. II. Series.
[DNLM: 1. Tyrosine—metabolism. 2. Cell Transformation. Neoplastic. QU 60 K293t 1994]
RC268.5.K45 1994
616.99'4071—dc20
DNLM/DLC 94-33516
for Library of Congress CIP

CONTENTS

Acknowledgments

In this monograph I have attempted to describe the inter-relationships between cell signaling and changes in the cytoskeleton in the context of neoplastic transformation. Any mistakes or misinterpretation of data are entirely my responsibility. Nevertheless I would like to thank those colleagues and friends whose support I have appreciated over the years, especially long-suffering past and present members of my laboratory, and former colleagues who played a role in the development of the ideas outlined in this monograph, whether they agreed with them or not. I am particularly indebted to the following people who have given help in the form of data, figures and constructive criticism: Simon Barry, David Critchley, Moira Elmore, George Felice, Peter Finan, Rita Horvath, John Lackie, Nick Matthews, Milan Nermut, and Lynn Wilson. Finally I would like to express my profound gratitude to Ann, Andrew and Ian for putting up with me during this project.

CHAPTER 1

INTRODUCTION

Over the last 20 years or so advances in molecular biology have led to many insights into the cell biology and biochemistry of cancer. Although a genetic component in cancer has been recognized for a number of years, the genes whose activities give rise to cancer were identified at the molecular level only 14-16 years ago, and even so it is becoming clear that many genes can cause cancer if they are expressed in an aberrant manner.

At present more is known about the genes which can cause cancer, or cause neoplastic transformation in vitro, than about the underlying biochemical mechanism by which they cause transformation-related changes within cells. A great deal of data has been amassed regarding the molecular biology of neoplasia in vivo and about the expression of genes which lead to transformation in a variety of cell culture and animal model systems, however the biochemical pathways which become deregulated due to the presence of these gene products are only just becoming understood. Clues about the biochemistry and cell biology of oncogenesis have gradually emerged resulting from the identification of the protein products of oncogenes as protein kinases, guanine nucleotide-binding proteins or nuclear factors; however despite this knowledge the mechanism by which these proteins functionally lead to the plethora of changes characteristic of tumor cells remains largely unknown.

This short monograph cannot cover all aspects of neoplastic transformation or tumorigenesis. For this reason I have chosen to concentrate on my own personal interests, which are admittedly heavily biased, and in the following chapters I will therefore describe the involvement of the cytoskeleton in the cellular changes manifested by transformed cells, and how changes in the contractile network of cells might lead to the transformed phenotype (or at least certain aspects of this). As I will develop in later chapters, common features of neoplastic transformation (in addition to unrestricted proliferation) are changes in the morphology, ultrastructure and surface architecture of cells including increased membrane protrusions, cell motility and a loss of the phenomenon known as contact inhibition of locomotion, where cells tend to respect each other's boundaries. All these events are controlled by the filamentous network within cells called the cytoskeleton. Within all cells the cytoskeleton comprises three major filamentous classes: microtubules, intermediate filaments and microfilaments. Many of the components of each of these systems have been identified, and using skeletal muscle as a model many of the biochemical interactions between proteins within the same filamentous system and between different filamentous systems have been characterized. Despite insights into the regulation of the cytoskeleton in normal cells, the reasons for the disorganization of the cytoskeleton in transformed cells or cells derived from tumors is generally unknown and, as I have already pointed out, the deregulation of cytoskeletal structures under pathological conditions such as tumorigenesis have not yet been totally elucidated, nor is it known how the cytoskeletal changes relate to other aspects of the neoplastic phenotype. In the following chapters I have attempted to expand on this and review the data suggesting that changes in the cytoskeleton are central to the maintenance of the transformed phenotype.

Many of the early oncogenes to be discovered belonged to the tyrosine kinase family of proteins, that is they had an enzyme activity which transferred a phosphate group from ATP to specific tyrosine residues in proteins. There is good reason to believe that tyrosine phosphorylation is an important feature of cellular growth control since increases in the levels of phosphotyrosine-containing proteins is a common feature of many transformed cells and many cells which respond to growth factors by proliferation. Intense investigation coupled with recent advances in molecular biology, developmental biology, cell biology and protein structure have given us much information on how many of these tyrosine kinase substrates interact to produce cellular or developmental changes. More recently still, the mechanism by which tyrosine kinases feed into other signaling pathways such as GTP-binding proteins has also started to be deciphered. In later chapters I have attempted to describe the interaction between different signaling molecules which contain recognition sites for tyrosine phosphorylated proteins and other molecules with respect to cytoskeletal changes. Finally, several tyrosine kinase-encoding genes have been implicated in the etiology of cancer in humans. There is direct evidence for the gene responsible for chronic myeloid leukemia being an activated tyrosine kinase, however despite a certain amount of correlative data there is at present little direct evidence for the involvement of activated tyrosine kinases in the major human cancers. Towards the end of this monograph I will describe the indirect evidence for the activation of proto-oncogenes such as *src* being involved in several human tumors.

The interested reader can find more information on some of the examples quoted above in the following list of general references. For specific details the reference list for each chapter should be consulted, where original sources have been used where practicable.

Further Reading

Alberts B, Bray D, Lewis J et al. Molecular Biology of the Cell. 3rd Edition. New York: Garland Publishing, 1994.

Bershadsky AD, Vasiliev JM. Cytoskeleton. New York: Plenum, 1990.

Bishop JM. Molecular themes in oncogenesis. Cell 1991; 64: 235-248.

Bishop JM, Varmus HE. Functions and origins of retroviral transforming genes. In: Weiss R, Teich N, Varmus H, Coffin J, eds. RNA Tumor Viruses. 2nd Edition. New York: Cold Spring Harbor Laboratory Press, 1984: 999-1108.

Bray D. Cell Movements. New York: Garland Publishing, 1992.

Evans CW. The Metastatic Cell. London: Chapman and Hall, 1991.

Kahn P, Graf T, eds. Oncogenes and Growth Control. Berlin: Springer, 1986.

Jove R, Hanafusa H. Cell transformation by the viral *src* oncogene. Annu Rev Cell Biol 1987; 3: 31-56.

Wyke JA. Viruses and Cancer. In : Franks LM, Teich NM, eds. Introduction to the Cellular and Molecular Biology of Cancer. 2nd Edition. Oxford: Oxford University Press, 1991: 203-229.

CHAPTER 2

The Discovery of Oncogenic Tyrosine Kinases

Current ideas on the role that oncogenes play in the development of human cancers stem from pioneering work on animal models first carried out over 80 years ago. These experiments were so important in our understanding of the etiology of cancer that it is worthwhile reviewing the history of the discovery of oncogenes in animals.

AVIAN SARCOMA VIRUSES

One of the most fruitful lines of research for understanding the mechanism resulting in neoplastic transformation has been the examination of the mode of action of various oncogene products. Oncogenes were originally discovered as genes encoded by a variety of different viruses whose presence was essential for cell transformation in vitro or tumor induction in vivo. These viral oncogenes (*v-onc*), evolved from closely related protooncogenes or cellular oncogene (*c-onc*) homologs which are present in normal cells. Oncogenes were first described as parts of the genome of acutely transforming retroviruses and these viruses usually contain one, or at most two, *v-onc* genes. Mutational analysis of RNA tumor viruses has demonstrated that the activity of a single gene, when present in the background of a retrovirus, is sufficient to induce neoplasia. Protooncogenes are evolutionarily conserved from simple organisms such as yeast to more complex organisms such as mammals, and this conservation has led to speculation that these genes are important in the regulation of normal cells. There is now growing evidence that inappropriate expression or activity of cellular protooncogenes can lead to neoplasia in mammalian cells.

A series of classic experiments was performed by Peyton Rous in the years 1907-1911. He excised the tissue of an apparently spontaneous chicken sarcoma and homogenized the material. After finely filtering the material to produce a cell-free extract he then injected this into another host bird. Within 14 days of injection this new chicken died of a sarcoma similar to that of the original bird. Early attempts to induce sarcomas in other species using the same tumor extract failed, showing that the agent which caused the tumors in chickens had a remarkable degree of host range specificity.[1] Over the next 30-40 years Rous established that the agent which induced the tumors in birds was an infective agent now termed Rous sarcoma virus. The new field of tumor virology was thus born, and although this was originally described in birds similar experiments have been performed in many other species. Molecular biology has now allowed fine detailed analysis of how these tumor viruses induce cancers in infected animals and later in this chapter I will discuss how the genes which cause cancer (oncogenes) were identified and characterized.

Although the early work of Rous clearly demonstrated the infectivity of the agent which caused tumors, it was not until the development of the electron microscope that the particulate nature of the infectious agent was demonstrated. Claude, Porter and Pickels[2] discovered small, electron dense particles in the cytoplasm of cells cultured from Rous sarcoma tumors and they suggested that the particles represented the virion particles of Rous sarcoma virus since the tumor-inducing activity of sarcoma extracts could be correlated with the number of particles present in such tumors. Some virus particles were also detected in apparently normal tissue suggesting that there might be a related family of endogenous nontransforming viruses which, when mutated, might be responsible for neoplastic diseases of chickens as leucosis or sarcomas. The possibility that viruses might be responsible for neoplastic disease has been a concern of the poultry industry for many years and certain leukemias such as Marek's disease have also been ascribed to viral infection.

The many years of successive culture of Rous sarcoma virus through cells has presumably selected a virus with particular potency for cell infection and growth, since almost all inoculated animals develop tumors even at low titers of inoculum.[3,4] The advent of tissue culture allowed the study of Rous sarcoma virus on cells in vitro, and it was found that cultured chicken fibroblasts infected by RSV became altered in appearance and histologically resembled sarcoma cells from tumor tissue. These altered cells occurred in discrete foci and the number of foci was proportional to the concentration of the infecting virus. Enumeration of the foci gives an accurate assay for the infectious titre of RSV preparations,[5,6] and this type of assay is still used today to assess viral transforming potential and multiplicity of infection.

RSV can induce tumors not only in chickens but also in many other birds.[7-10] Latterly other strains of RSV have been isolated which will efficiently infect mammalian cells and these strains can give rise to tumor induction in rats,[11] mice,[12,13] Syrian hamsters, guinea pigs,[14] dogs[15] and monkeys.[16]

RSV belongs to the retrovirus family, and its genome therefore consists of RNA rather than DNA.[17] Genetic analysis has shown that it contains the classical *gag*, *pol* and *env* genes encoding proteins involved in the assembly of the virion particle outer coat, the lipoprotein envelope and also generation of reverse transcriptase to transcribe the viral RNA into DNA for insertion into the host genome, but in addition contains a transforming (*src*) gene as depicted in Figure 2.1. The malignant changes induced by RSV have been shown to be the consequence of expression of the *src* gene because mutations or deletions of this region destroy its tumorigenic activity.[17-20] This is supported by the findings that if chickens are injected with RSV mutants containing deletions in the *src* gene no tumors occur within the normal time scale for tumorigenesis with wild-type RSV. Occasional tumors do arise in animals injected with *src* deletion mutants after a lag time of anything up to four months, and viruses which can be recovered from these tumors usually contain a functional *src* gene generated by a recombination of viral and cellular sequences.[21] These recovered transforming viruses invariably contain a genetically altered *src* gene. Thus deletion of a *src* gene results in loss of transforming ability and regeneration of a functional *src* gene rescues the transforming phenotype of the virus. The level of *v-src* protein in RSV-transformed cells can be up to 50 times higher than the levels of the endogenous normal cellular protein encoded by *c-src*. This led to early suggestions that overexpression of the *src* gene product leads to transformation.

THE *V-SRC* ONCOGENE

The Origin of a Transforming Avian Sarcoma Virus

Nontransforming retroviruses are endogenous components of the genome of birds and mammals, and in order for a nontransforming virus to replicate after infection the viral RNA within the cell has to be converted to DNA by the activity of the viral reverse transcriptase. This resulting DNA provirus is then inserted apparently at random into the host genome. It is possible that occasionally there is a fusion between viral and cellular sequences to produce a viral/cellular fusion product. After processing, the resulting product would contain viral 5' sequences, viral 3' sequences and also sequences derived from cellular exons. If the genes encoding the virion particles are conserved then multiple copies of the fused RNA could then be packaged as normal. If the captured cellular sequences are potentially oncogenic, during the next round of infection the oncogenic sequences would be under the control of viral promoters and so the protooncogenic gene will be aberrantly expressed. In most transforming avian retroviruses this recombination event is accompanied by an alteration of the protooncogenic sequences which might be mutagenesis, truncation, deletion, substitution or a combination of these. The integration and subsequent alteration of *c-src* in an avian retrovirus is depicted in Figure 2.2.

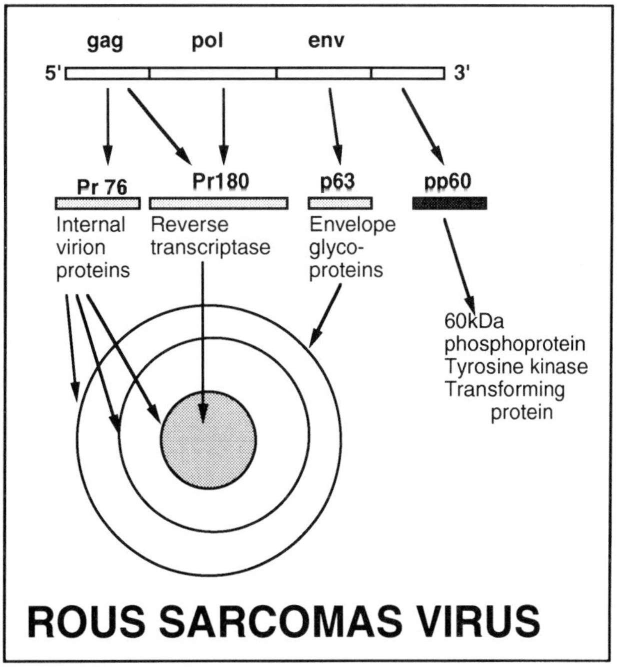

Fig. 2.1. Rous sarcoma virus genome. The RSV genome encodes four genes: three structural genes, gag, pol *and* env *encoding virion coat proteins, reverse transcriptase and envelope glycoproteins respectively plus the* src *gene which encodes* $pp60^{v\text{-}src}$.

Several strains of RSV have been isolated which all have very similar *src* genes but which are placed in slightly different positions in the RSV genome. The Schmidt-Rupin and Prague strains are the archetypal viruses which can replicate due to preservation of a complete genome encoding functional virion particle proteins (see Fig. 2.1), whereas other strains need a helper virus in order to replicate. In these replicative viruses the *src* gene is placed at the 3' end of the viral genome. Cloning and sequencing of the normal chicken *c-src* gene has shown that all *v-src* genes preserve the same *c-src* initiation codon, therefore all *v-src* proteins contain virtually identical amino-terminal amino-acid sequences.[22-24] At the 5' end, there is divergence between the viral and cellular genes, possibly due to two recombination events replacing the last 57 base pairs of *c-src* with 31 base pairs located downstream of the *c-src* termination codon plus another 31 bases of viral flanking region. This results in the substitution of the carboxy-terminal 19 amino acids of *c-src* with a new sequence of 12 amino acids which then comprise *v-src*. Thus the *c-src* gene encodes a protein of 533 amino acids whereas the *v-src* gene encodes a protein of

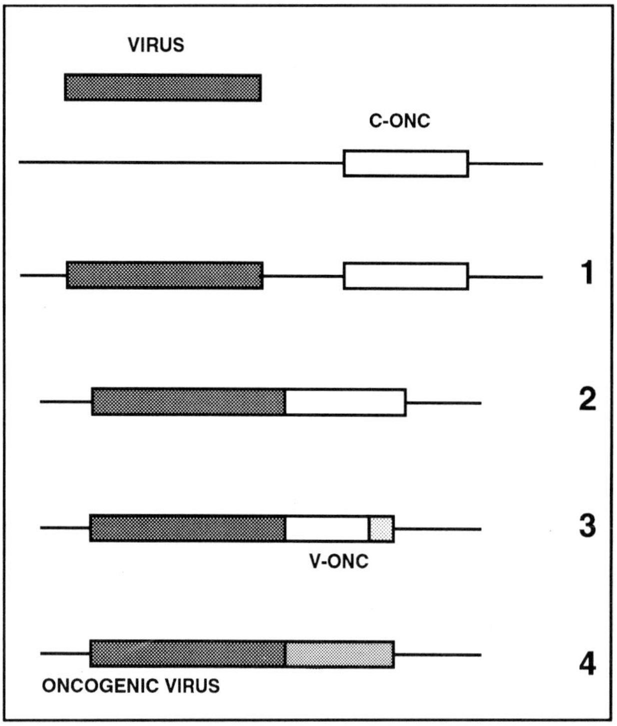

Fig. 2.2. Integration of a cellular protooncogene into an avian retrovirus to produce an acutely transforming retrovirus, using RSV as an example.

1. *Random integration of proviral DNA into the host genome, occasionally close to a cellular protooncogene (c-onc).*
2. *A recombination event which excises intervening sequences between proviral DNA and c-onc DNA places c-onc in the same reading frame as proviral DNA.*
3. *In addition there are usually changes in the new proviral c-onc fusion protein, e.g., substitution or deletion, resulting in new proviral DNA containing an activated oncogene. In the case of c-src there is a deletion then partial replacement at the 3' end of the gene.*
4. *Production of new oncogenic retroviral RNA genome.*

526 amino acids, with severe consequences for its enzyme regulation as will be described chapter 5.[25]

Biochemical Activity of the *SRC* Gene Product

The *v-src* gene encodes a protein containing 526 amino acids designated pp60$^{v\text{-}src}$.[26] Soon after its discovery this protein was shown to have protein kinase activity, but more significantly several groups showed independently that the activity was that of a tyrosine-specific protein kinase.[27-32] Since the vast majority of phosphorylation which normally occurs within cells is on serine or threonine it was suggested that tyrosine-specific phosphorylation of cellular substrates was the mechanism by which RSV induces transformation and evidence for this will be discussed in further chapters.[33,34] Subsequently pp60$^{v\text{-}src}$ has been shown to contain several discrete domains which regulate its function and I shall discuss the relevance of this later in this chapter and also in chapter 5.

Synthesis and Localization of pp60$^{v\text{-}Src}$

pp60$^{v\text{-}src}$ is synthesized on free polyribosomes from 21S viral mRNA in the cell cytoplasm.[35,36] The initiator methionine residue of the nascent polypeptide is then cleaved and acylation of the protein occurs immediately after translation (Fig. 2.3). The first 14 amino acids of pp60$^{v\text{-}src}$ contain a recognition site for the addition of a 14 carbon lipid, myristic acid, to the new N-terminal glycine.[37-41] This addition appears to be essential for subsequent localization of this protein to the plasma membrane, since mutants with N-terminal deletions which are not myristoylated are cytosolic. Whilst covalent lipid might be necessary for membrane localization, this may not be sufficient for membrane attachment since at least one temperature-sensitive mutant is *ts* for membrane localization but is constitutively myristoylated with the result that at the restrictive temperature these *ts*RSV-transformed cells contain myristoylated pp60$^{v\text{-}src}$ which is cytoplasmic.[42] Although many myristoylated proteins within cells are associated with membranes there are some exceptions, for example the catalytic subunit of cAMP-dependent protein kinase is myristoylated but appears to be a cytosolic protein, so myristoylation is not necessarily a membrane localization signal for all proteins.[43]

Immunoprecipitation of newly synthesized pp60$^{v\text{-}src}$ from transformed cells has shown that immediately or soon after synthesis pp60$^{v\text{-}src}$ associates with two cellular proteins of 50kDa and 90kDa (p50 and p90).[44-49] These proteins were shown to be distinct from pp60$^{v\text{-}src}$ since they exhibited different peptide profiles after proteolysis, and they form a heterotrimeric complex with pp60$^{v\text{-}src}$ with a stoichiometry of 1:1:1. Sedimentation analysis of pp60$^{v\text{-}src}$ from transformed cells has shown that pp60$^{v\text{-}src}$ exists in two forms within cells: a monomeric form and a complexed form; however greater than 90% of the newly synthesized pp60$^{v\text{-}src}$ associates with p50 and p90 such that at steady state about 5% of the total pp60$^{v\text{-}src}$ within cells is complexed. The complex is extremely stable, being resistant to high salt concentrations, chelating agents or reducing agents.[44] The p50 and p90 proteins do not need to be newly synthesized in order to bind pp60$^{v\text{-}src}$, suggesting that they can be re-utilized for complex formation. Pulse-chase experiments have demonstrated that this cytoplasmic complex has a half-life of less than 15 minutes which is similar to that of the translocation of pp60$^{v\text{-}src}$ to the plasma membrane, suggesting that this complex targets pp60$^{v\text{-}src}$ to the plasma membrane. The demonstration that the complexed pp60$^{v\text{-}src}$ is not associated with the plasma membrane whereas the uncomplexed pp60$^{v\text{-}src}$ is tightly associated with the plasma membrane is consistent with this transcytoplasmic transportation function for p50 and p90.[48,50] p90 is an abundant cellular protein which appears to be cytoplasmic since it is freely soluble when cells are fractionated into cytosol/membranes or into detergent soluble/insoluble (cytoskeletal) fractions.[51] It has been established that p90 is a heat shock protein whose expression is increased when cells are subjected to stress. p50 is a less abundant phosphoprotein which in nontransformed

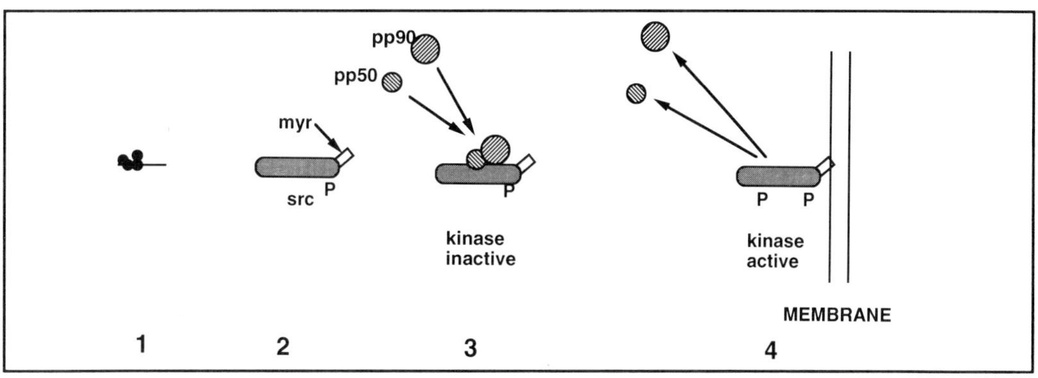

Fig. 2.3. Synthesis, processing and localization of pp60$^{v\text{-}src}$. 1; pp60$^{v\text{-}src}$ is synthesized on free cytosolic polysomes from proviral mRNA. 2; Immediately after synthesis the initiator methionine residue is removed and the new N-terminal glycine is myristoylated. 3; p50 and p90 associate with pp60$^{v\text{-}src}$, the enzyme is phosphorylated on serine 17, and this complex translocates to the plasma membrane with a half-life of 15-20 minutes. While in this complex pp60$^{v\text{-}src}$ has no kinase activity. 4; Upon reaching the membrane p50 and p90 dissociate, pp60$^{v\text{-}src}$ gains tyrosine kinase activity and autophosphorylation on tyrosine 416 occurs. Only membrane-associated pp60$^{v\text{-}src}$ is transforming.

cells contains predominantly phosphoserine, however in transformed cells the p50 population which associates with pp60$^{v\text{-}src}$ contains phosphotyrosine in addition to phosphoserine.[52,53] It is unlikely that this tyrosine phosphorylation is important in pp60$^{v\text{-}src}$ binding since p50 is not tyrosine phosphorylated in transformed rodent cells although complex formation occurs efficiently.[46] Thus there is no evidence that tyrosine phosphorylation of p50 has any functional consequence for complex formation or translocation. Mutant viruses have been used to map the regions of pp60$^{v\text{-}src}$ to which p50 and p90 bind. Deletion analysis has shown that pp60$^{v\text{-}src}$ and p90 do not bind to the amino terminal part of pp60$^{v\text{-}src}$, and further analysis has localized the complex formation of p50 and p90 to the region of pp60$^{v\text{-}src}$ between amino acids 81 and 173. Antibody analysis has been less specific in suggesting that the carboxy-terminal half of the protein is important for complex formation.[54]

Immunoprecipitates of complexed pp60$^{v\text{-}src}$ have little or no kinase activity as assessed by autophosphorylation or phosphorylation of exogenous substrates, perhaps due to masking of the enzyme site or regulatory region by the two cellular proteins,[44,48] however there is little direct evidence for this. Recently the use of yeast expression systems has led to the surprising discovery that contrary to predictions the binding of HSP90 to pp60$^{v\text{-}src}$ results in stimulation of kinase activity.[55] It may be that p50 is also necessary for kinase inhibition, however these results indicate that HSP90 does not simply inhibit pp60$^{v\text{-}src}$ tyrosine kinase activity during transit to the plasma membrane.

Pulse-chase experiments have shown that there is a transcytoplasmic movement of pp60$^{v\text{-}src}$ toward the plasma membrane and within 15 minutes pp60$^{v\text{-}src}$ becomes localized at the plasma membrane.[49,56] It then dissociates from the complex and binds to the plasma membrane. This is followed by phosphorylation of the major autophosphorylation site (tyrosine 416) and activation of the tyrosine protein kinase function of pp60$^{v\text{-}src}$. The half life of pp60$^{v\text{-}src}$ under normal physiological conditions is about 60 minutes and temperature-sensitive mutant pp60^{src} proteins have similar half-lives to the wild type, irrespective of whether the cells are maintained at the nonpermissive or permissive temperature.[57]

Functional Domains of pp60$^{v\text{-}Src}$

The *src* gene product can be divided into a number of domains based on several criteria: examples are amino acid sequence homologies, proteolysis and functional examination of peptide fragments and analysis of deletions or specific mutations. Amino acid sequence analysis has led to the discovery that several regions of the *src* gene exhibit homologies with other proteins (Fig. 2.4). The region containing the first 14 amino acids is conserved for many members of the *src* family which contain myristic acid covalently linked to the N-terminal glycine residue.[58] This sequence contains a recognition motif for the enzyme which catalyses myristoylation, myristyl CoA transferase.[59] Other regions have been identified by sequence analysis which have also been found in a variety of other signaling molecules: Src homology domains 2 and 3 (SH2 and SH3). The structure and function of these domains will be described in detail in chapter 5. SH2 domains are responsible for the recognition and association of proteins with tyrosine phosphorylation sites of other molecules and thus direct complex formation between signaling molecules.[60-62] SH2 domains also appear to be responsible for the interaction of pp60$^{v\text{-}src}$ with the cytoskeleton as will be described, however in other proteins it appears to be the SH3 domains which confer cytoskeletal association.[63] The region spanning the SH2 and SH3 domains is a regulatory region which also plays a role in conferring substrate specificity.[64] The carboxy-terminal half of pp60^{src} contains the catalytic domain of the protein and the ATP-binding site, and this region has a high degree of homology with all other tyrosine kinases, including those of the growth factor receptor family and other intracellular tyrosine kinases such as *syk* and JAK.[65-67]

A great deal of useful information about the various domains on pp60$^{v\text{-}src}$ has been gained by the study of proteolytic fragments of pp60$^{v\text{-}src}$, but even more useful has been the use of deletion mutants for these studies. Using these methods of investigation it has been shown that a carboxy terminal proteolytic fragment of 30kDa retains protein-tyrosine kinase activity, reflecting the significant homology of this region to other tyrosine and serine protein kinases. Mutation analysis has revealed that the amino terminal 8 to 15 kDa are implicated in attachment to the plasma membrane.[68,69,31] This amino terminal is not appreciably hydrophobic but myristic acid is covalently attached through an amide linkage to the N-terminal glycine residue,[39-41] and this feature seems crucial (but not sufficient) for membrane localization. A region in the amino-terminal half of pp60$^{v\text{-}src}$ with minimum boundaries from amino acids 81 and 169 has an ill-defined modulatory role in pp60$^{v\text{-}src}$ function. Many mutants with lesions in this region possess apparently normal kinase activity, but induce an aberrant morphology in transformed cells, possibly reflecting an impaired or altered substrate specificity.[70,71] Recently Fukui et al[73] have mapped the domain involved in cytoskeletal association to residues 149-169.

The binding of pp60$^{v\text{-}src}$ to plasma membranes is saturable and dependent on the presence of myristate at the amino terminus of pp60$^{v\text{-}src}$. Nonmyristoylated *src* proteins are unable to bind to the plasma membrane and are unable to induce morphological alteration and anchorage-independent growth in soft-agar suspension which are both common characteristics of the transformed phenotype. There is also a good correlation between the association of pp60$^{v\text{-}src}$ with the detergent-resistant cytoskeleton and cell transformation, and the cytoskeletal proteins with which pp60$^{v\text{-}src}$ interacts might also be associated with the plasma membrane since pp60$^{v\text{-}src}$ is localized to the specialized membrane-associated adhesion structures known as adhesion plaques.[74,75] An indication that there might be a specific membrane receptor for pp60$^{v\text{-}src}$ comes from the findings that prior treatment of cell membranes with heat or trypsin greatly decreases subsequent binding of pp60$^{v\text{-}src}$. This membrane binding of pp60$^{v\text{-}src}$ can be competed by a myristoylated peptide containing the first 11 amino acids of the mature *src* sequence, but not by non-myristoylated *src* peptide or

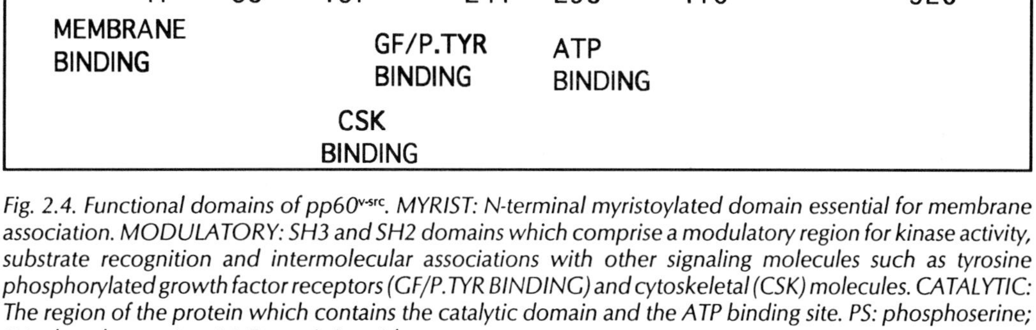

Fig. 2.4. Functional domains of pp60$^{v\text{-}src}$. MYRIST: N-terminal myristoylated domain essential for membrane association. MODULATORY: SH3 and SH2 domains which comprise a modulatory region for kinase activity, substrate recognition and intermolecular associations with other signaling molecules such as tyrosine phosphorylated growth factor receptors (GF/P.TYR BINDING) and cytoskeletal (CSK) molecules. CATALYTIC: The region of the protein which contains the catalytic domain and the ATP binding site. PS: phosphoserine; PY: phosphotyrosine; MYR: myristic acid.

other myristoylated peptides again suggesting the existence of a *src* receptor in cell membranes.[76] Resh and Ling have described a 32kDa membrane protein that binds to the myristoylated amino-terminal sequence of pp60$^{v\text{-}src}$ and has the characteristics of a receptor for pp60$^{v\text{-}src}$.[77] However later experiments have identified the 32kDa protein as the mitochondrial ADP/ATP carrier, and whilst this receptor bound N-terminally myristoylated *src* peptide it could not bind pp60^{src} polypeptide, therefore the relevance of this is unclear and it is unlikely that this is the protein which anchors pp60$^{v\text{-}src}$ top the plasma membrane.[78]

NONTRANSFORMING EFFECTS OF PP60$^{V\text{-}SRC}$ ON CELLS

EFFECTS ON DIFFERENTIATION

The effect of *src* on cells with respect to neoplastic transformation will be covered in the next chapter, however oncogenic tyrosine kinases can have an effect on cells which results in changes in differentiation patterns rather than proliferation. Some cancers are the result of the inability of cells to undergo terminal differentiation and so changes in the differentiation program of cells are crucially important in the ability of oncogenes to induce cancer in vivo.

pp60$^{v\text{-}src}$ has a morphoregulatory role when expressed at low levels in epithelial cells such as MDCK cells and this may mimic events controlled by an endogenous related cellular tyrosine kinase.[79,80] The effect of the expression of the *v-src* gene on the cellular differentiation program is often cell type specific. In myoblasts and some neuron- related cells the expression of differentiated cell functions is suppressed. An example of this is described by Falcone et al[81] who showed that when embryonic quail myoblasts were infected with Rous sarcoma virus, Fujinami sarcoma virus or avian erythroblastosis virus the cells took on a transformed phenotype including increased proliferation rates, the ability to grow in semi-solid media and morphological alterations. These cell populations failed to fuse into multinucleate myotubes, thus showing a defect in terminal differentiation. The role of pp60$^{v\text{-}src}$ in this inhibitory process was confirmed by the demonstration that infecting the cells with a temperature-sensitive mutant of RSV resulted in a temperature-sensitive inhibition of terminal differentiation. In addition when myoblasts were transformed by RSV a higher amount of hyaluronic acid was synthesized in conjunction with an inability to fuse.[82] In contrast, *v-src* can stimulate the differentiation of pheochromocytoma cells[83] but will also inhibit glial cell differentiation.[84] When *fps* is introduced into chicken myeloid cells there is no difference in the ability of bone marrow cells to differentiate into macrophages in response to growth factors, however unlike the normal counterparts *fps*-infected bone marrow cells can also differentiate in a growth factor-independent manner indicating that there is an inter-relationship between oncogenes and growth factors in differentiation as well as in proliferation.[85]

pp60$^{v\text{-}src}$ expression can also inhibit pigment production, differentiation and junctional communication of chicken embryo lens cells.[86] Since intercellular communication appears important in the function of normal epithelial cells, some responses of cells to oncogenic tyrosine kinases which are thought of as transformation parameters might just as easily be the result of inhibition of differentiation. For example structural modification of gap junctions and a reduction in junctional communication is an early event after activation of pp60$^{v\text{-}src}$.[87] Expression of the *src* gene can also induce changes in nonjunctional membrane transport systems, for example transformed cells are less sensitive to density-dependent inhibition of phosphate transport compared to nontransformed cells, and the expression of pp60$^{v\text{-}src}$ can also increase the transport of potassium, amino acids, and glucose across the plasma membrane in transformed cells.[88] Yet another example of the regulation of differentiation by tyrosine kinases is found in their effect on the maturation of frog eggs. Microinjection of pp60$^{v\text{-}src}$ into

Xenopus laevis oocytes resulted in an increase in the rate of progesterone- or insulin-induced meiotic maturation.[89] Thus tyrosine kinases are important regulatory molecules from the earliest stages of development.

Inhibition of *src* Transformation in Embryos

I have already described the induction of tumors in mature chickens by RSV and the neoplastic transformation of cells in vitro. It is possible that multiple factors might also be involved in tumor formation in vivo, for example the induction and progression of papillomas in mouse skin needs the cooperation of two oncogenes or two factors, one a promoter and the other an inducer. In order to generate stable transformation of mammalian cells in vitro the cells first have to be immortalized before full transformation can occur.[90-95] This multifactorial origin of tumorigenesis has also been studied in vivo using chick embryos as a model system. Bissell and co-workers have investigated the ability of RSV to induce tumors in chick embryos or newly-hatched birds rather that mature animals. They found that embryos and newly hatched birds were refractory for general tumorigenesis, and that after injection with RSV tumors developed only at the site of injection despite the fact that the virus was circulating in the bloodstream and that $pp60^{v\text{-}src}$ was present in the tissues. They also showed that when fibroblasts from these animals were placed in culture, after a 24 hour lag period all the cells spontaneously transformed, therefore demonstrating that the cells have the potential to transform but that there was something about the embryonic environment which suppressed tumorigenesis. Furthermore if a wound was inflicted on the newly hatched, RSV-injected bird at a site distant from the injection site a tumor formed where wounding occurred. This work demonstrated that factors in addition to $pp60^{v\text{-}src}$ were involved in tumorigenicity and suggested that generation of growth factors was a necessary component of tumor formation, particularly in embryos or neonates.[96-98] Further studies showed that if *v-src* was introduced into selected embryonic cells by retroviral transduction clones of cells containing *v-src* could be traced within a normal embryonic background. These cells grew normally within the developing embryo except for endothelial cells which produced tumors, however neoplastic transformation occurred when these cells were placed into culture.[99] Bissell and co-workers have concluded that cells within an embryo are subject to strict homeostatic controls which are lost when cells are put into culture. The regulatory factors or controls have yet to be identified, however there is a growing knowledge about the role of tumor suppressor genes in cell proliferation and the embryonic factors may be related to the regulation of such genes.

Transgenic Animal Models

Since $pp60^{c\text{-}src}$ is highly expressed in platelets and neurones, it might be expected that it plays a crucial role in tissue-specific development or cell function. In order to examine the role of *c-src* in normal animals, transgenic homozygous mice have been developed in which the *c-src* gene has been inactivated by targeted gene disruption. When these animals were examined for developmental defects or changes in cell function, somewhat surprisingly there was little apparent pathology. In particular homozygous null *src* mice had no nervous system defects, platelet dysfunction or wound healing problems. The only phenotype associated with *src* knockout was osteopetrosis, which is characterized by a decrease in bone resorption. Osteoclasts are the major cell type which is responsible for bone resorption, however since osteoclasts were still present in these mice the defect must be in osteoclast function rather than generation, although the exact functional defect has yet to be defined.[100] By contrast knockouts of other tyrosine kinases do result in specific phenotypes, for example mice in which the *src*-related tyrosine kinase *fyn* has been deleted have impaired long term potentiation, spatial learning and hippocampal development.[101] Further use of transgenic animals will be a powerful tool to investigate normal cellular tyrosine kinase function and help define the role of these proteins in normal and pathological conditions. The following chapters will describe the effect of the activated *src* gene product, $pp60^{v\text{-}src}$, on cell function.

References

1. Rous P. A sarcoma of the fowl transmissible by an agent separable from the tumor cells. J Exp Med 1911; 13: 397-411.
2. Claude A, Porter KR, Pickels EG. Electron microscope study of chicken tumor cells. Cancer Res 1947; 7: 421-430.
3. Epstein MA. Intracellular identification of the Rous virus. Nature (London) 1956; 178: 45-46.
4. Epstein MA, Holt SJ. Observations on the Rous virus: integrated electron microscopical and cytochemical studies of fluorocarbon prepared material. Brit J Cancer 158; 12: 363-369.
5. Halberstaedtler L, Doljanski L, Tenenbaum E. Experiments on the cancerization of cells in vitro by means of Rous sarcoma virus. Brit J Exp Path 1941; 22: 179-187.
6. Doljanski L, Tenenbaum E. Studies on Rous sarcoma cells culivated in vitro I. Cellular composition of pure cultures of Rous sarcoma cells. Cancer Res 1942; 2: 776-785.
7. Andrewes CH. Some properties of immune sera active against fowl-tumor viruses. J Path Bact 1932; 35: 243-249.
8. Duran-Reynals F. The reciprocal infection of ducks and chickens with tumor-inducing viruses. Cancer Research 1942; 2: 243-369.
9. Duran-Reynals F. The infection of turkeys and guinea fowl by the Rous sarcoma virus and the accompanying variations of the virus. Cancer Res 1943; 3: 569-577.
10. Borges PRF, Duran-Reynals F. On the induction of malignant tumors in pigeons by a chicken sarcoma virus after previous adaptation of the virus to ducks. Cancer Res 1952; 12: 55-58.
11. Svet-Moldavsky GJ. Sarcoma in albino rats treated during the embryonic stage with Rous virus. Nature (London) 1958 182: 1452-1453.
12. Svoboda J. Presence of chicken tumor virus in the sarcoma of adult rats inoculated after birth with Rous sarcoma tissue. Nature (London) 1960; 186: 980-982.
13. Ahlstrom CG, Kato R, Levan A. Rous sarcoma in chinese hamsters. Science 1964; 144: 1232-1233.
14. Ahlstrom CG, Forsby N. Sarcomas in hamsters after injection with Rous chicken tumor material. J Exp Med 1962; 115: 839-852.
15. Rabotti GF, Bucciarelli E, Dalton AJ. Presence of particles with the morphology of viruses of the avian leukosis complex in meningeal tumors induced in dogs by Rous sarcoma virus. Virol 1966; 29: 684-686.
16. Munroe JS, Windle WF. Tumors induced in primates by chicken sarcoma virus. Science 1963; 140: 1415-1416.
17. Hanafusa H. Cell transformation by RNA tumor viruses. In: Fraenkel-Conrat H, Wagner RR, eds. Comprehensive Virology. New York: Plenum, 1977: 401-483.
18. Yacuik P, Shalloway D. Features of the $pp60^{src}$ carboxyl terminus that are required for transformation. Mol Cell Biol 1986; 6: 2807-2819.
19. Hanafusa H, Hanafusa T, Rubin H. The defectiveness of Rous sarcoma virus. Proc Natl Acad Sci USA 1963; 49: 572-590.
20. Teich N, Wyke J, Mak T et al. Pathogenesis of retrovirus-induced disease. In: Weiss R, Teich N, Varmus H, Coffin J, eds. RNA Tumor Viruses. 2nd Edition. New York: Cold Spring Harbor Laboratory Press, 1984: 785-998.
21. Hanafusa H, Halpern CC, Buchhagen DL et al. Recovery of avian sarcoma virus from tumors induced by transformation-defective mutants. J Exp Med 1977; 146: 1735-1747.
22. Linial M, Blair D. Genetics of retroviruses. In: Weiss R, Teich N, Varmus H, Coffin J, eds. RNA Tumor Viruses. 2nd Edition. New York: Cold Spring Harbor Laboratory Press, 1984: 649-703.
23. Takeya T, Hanafusa H. Structure and sequence of the cellular gene homologous to the RSV *src* gene and the mechanism for generating the transforming virus. Cell 1983; 32: 881-890.

24. Opperman H, Levinson AD, Varmus HE et al. Uninfected vertebrate cells contain a protein that is closely related to the product of avian sarcoma virus transforming gene (*src*). Proc Natl Acad Sci USA 1979; 76: 1804-1808.
25. Hunter T. A tail of two src's: mutatis mutandis. Cell 1987; 49: 1-4.
26. Jove R, Hanafusa H. Cell transformation by the viral *src* oncogene. Ann Rev Cell Biol 1987; 3: 31-56.
27. Collett MS, Erikson RL. Protein kinase activity associated with the avian sarcoma virus *src* gene product. Proc Natl Acad Sci USA 1978; 75: 2021-2024.
28. Levinson AD, Opperman H, Levotow L et al. Evidence that the transforming gene of avian sarcoma virus encodes a protein kinase associated with a phosphoprotein. Cell 1978; 15: 561-572.
29. Hunter T, Sefton BM. Transforming gene product of Rous sarcoma virus phosphorylates tyrosine. Proc Natl Acad Sci USA 1980; 77: 1311-1315.
30. Collett MS, Purchio AF, Erikson RL. Avian sarcoma virus transforming protein, pp60^{src}, shows protein kinase activity specific for tyrosine. Nature (London) 1980; 285: 167-169.
31. Levinson AD, Opperman H, Varmus HE et al. The purified protein product of the transforming gene of avian sarcoma virus phosphorylates tyrosine. J Biol Chem 1980; 255: 11973-11980.
32. Gilmer TM, Erikson RL. Rous sarcoma virus transforming protein, pp60^{src}, expressed in E. coli functions as a protein kinase. Nature (London) 1981; 294: 771-773.
33. Sefton BM, Hunter T, Beemon K et al. Evidence that the phosphorylation of tyrosine is essential for cellular transformation by Rous sarcoma virus. Cell 1980; 20: 807-816.
34. Wyke JA, Stoker A. Genetic analysis of the form and function of the viral *src* gene product. Biochim Biophys Acta 1987; 907: 47-69.
35. Lee JS, Varmus HE, Bishop JM. Virus-specific messenger RNAs in permissive cells infected by avian sarcoma virus. J Biol Chem 1979; 254: 8015-8022.
36. Purchio AF, Jovanovich S, and Erikson RL. Sites of synthesis of viral proteins in avian sarcoma virus-infected chicken cells. J Virol 1980; 35: 629-636.
37. Sefton BM, Trowbridge IS, Cooper JA et al. The transforming proteins of Rous sarcoma virus, Harvey sarcoma virus and Abelson virus contain tightly bound lipid. Cell 1982; 31: 465-474.
38. Garber EA, Cross FR, Hanafusa H. Processing of p60^{src} to its myristylated membrane-bound form. Mol Cell Biol 1985; 5: 2781-2788.
39. Schultz AM, Henderson LE, Oroszlan S et al. Amino terminal myristylation of the protein kinase p60^{src}, a retroviral transforming protein. Science 1985; 227: 427-429.
40. Buss, JE, Kamps MP, Sefton BM. Myristic acid is attached to the transforming protein of Rous sarcoma virus during or immediately after synthesis and is present in both soluble and membrane-bound forms of the protein. Mol Cell Biol 1984; 4: 2697-2704.
41. Buss JE, Sefton BM. Myristic acid, a rare fatty acid, is the lipid attached to the transforming protein of Rous sarcoma virus and its cellular homolog. J Virol 1985; 53: 7-12.
42. Stoker AW, Kellie S, Wyke JA. Intracellular localization and processing of pp60$^{v\text{-}src}$ proteins expressed by two distinct mutants of Rous sarcoma virus. J Virol 1986; 58: 876-883.
43. Carr SA, Biemann K, Shoji S et al. N-tetradecanoyl is the NH2 blocking group of the catalytic subunit of cyclic AMP dependent protein kinase from bovine cardiac muscle. Proc Natl Acad Sci USA 1982; 79: 6128-6131.
44. Brugge JS, Erikson E, Erikson RL. The specific interaction of the Rous sarcoma virus transforming protein, pp60$^{v\text{-}src}$, with two cellular proteins. Cell 1981; 25: 363-372.
45. Opperman H, Levinson W, Bishop JM. A cellular protein which associates with a transforming protein of Rous sarcoma virus is a heat-shock protein. Proc Natl Acad Sci USA 1981; 78: 1067-1071.
46. Opperman H, Levinson W, Levintow L et al. Two cellular proteins that immunoprecipitate with the transforming protein of Rous sarcoma virus. Virology 1981;

113:736-751.

47. Lipsich LA, Cutt J, Brugge JS. Association of the transforming proteins of Rous, Fujinami and Y73 avaian sarcoma viruses with the same two cellular proteins. Mol Cell Biol 1982; 2: 875-880.
48. Brugge JS, Yonemoto W, Darrow D. Interaction between the Rous sarcoma virus transforming protein and two cellular phosphoproteins: Analysis of the turnover and distribution of this complex. Mol Cell Biol 1983; 3: 9-19.
49. Levinson AD, Courtneidge SA, Bishop JM. Structural and functional domains of the Rous sarcoma virus transforming protein ($pp60^{src}$). Proc Natl Acad Sci USA 1981; 78: 1624-1628.
50. Courtneidge SA, Levinson AD, Bishop JM. The protein encoded by the transforming protein of Rous sarcoma virus ($pp60^{v-src}$) and homologous protein in normal cells ($pp60^{proto-src}$) are associated with the plasma membrane. Proc Natl Acad Sci USA 1980; 77: 3783-3787.
51. Yonemoto W, Lipsich L, Darrow D et al. An analysis of the interaction of the Rous sarcoma virus transforming protein. In: Schlesinger MJ, Ashburner M, Tissieres A, eds. Heat shock proteins: from bacteria to man. New York: Cold Spring Harbor Laboratory, 1982: 289-298.
52. Brugge JS, Darrow D. Rous sarcoma virus-induced phosphorylation of a 50,000 molecular weight cellular protein. Nature (London) 1982; 295: 250-253.
53. Gilmore T, Radke K, Martin GS. Tyrosine phosphorylation of a 50K cellular polypeptide associated with the Rous sarcoma virus transforming protein, $pp60^{src}$. Mol Cell Biol 1982; 2: 199-206.
54. Tamura T, Bauer H, Birr C et al. Antibodies against synthetic peptides as a tool for functional analysis of the transforming protein $pp60^{src}$. Cell 1983; 34: 587-596.
55. Xu Y, Lindquist S. Heat-shock protein hsp90 governs the activity of $pp60^{v-src}$ kinase. Proc Natl Acad Sci USA 1993; 90: 7074-7078.
56. Courtneidge SA, Bishop JM. Transit of $pp60^{v-src}$ to the plasma membrane. Proc Natl Acad Sci USA 1982; 79: 7117-7121.
57. Zemeicki A. Characterization of the monomeric and complex-associated forms of the gag-onc fusion proteins of three isolates of feline sarcoma virus: phosphorylation, kinase activity, acylation and complex formation. Virol 1986; 151: 265-273.
58. Cross FR, Garber EA, Pellman D et al. A short sequence in the $pp60^{src}$ N terminus is required for myristylation and membrane association and for transformation. Mol Cell Biol 1984; 4: 1834-1842.
59. Glover CJ, Goddard C, Felsted RL. N-myristoylation of $p60^{src}$. Identification of a myristyoyl-CoA:glycylpeptide N-myristoyltransferase in rat tissue. Biochem J 1988; 250: 485-491.
60. Cantley LC, Auger KR, Carpenter C et al. Oncogenes and signal transduction. Cell 1991; 64: 281-302.
61. Panayotou G, Waterfield MDO. The assembly of signaling complexes by receptor tyrosine kinases. BioEssays 1993; 15: 171-177.
62. Pawson T, Schlessinger J. SH2 and SH3 domains. Current Biology 1993; 3: 434-442.
63. Pawson T, Gish GD. SH2 and SH3 domains: from structure to function. Cell 1992; 71: 359-362.
64. Mayer BJ, Baltimore D. Signaling through SH2 and SH3 domains. Trends Cell Biol 1993; 3: 8-13
65. Hanks SK, Quinn AM, Hunter T. The protein kinase family: conserved features and deduced phylogeny of the catalytic domains. Science 1988; 241: 42-52.
66. Agarwal A, Salem P, Robbins KC. Involvement of $p72^{syk}$, a protein-tyrosine kinase, in Fcγ receptor signaling. J Biol Chem 1993; 268: 15900-15905.
67. Silvenoinnen O, Ihle JN, Schlessinger J et al. Interferon-induced nuclear signaling by Jak protein tyrosine kinase. Nature (London) 1993; 366: 583-585.
68. Krueger JG, Wang E, Garber EA, Goldberg AR. Subcellular localization of $pp60^{src}$ in RSV-transformed cells. Curr Topics Microbiol Immunol 1983; 107: 51-124.
69. Krueger JG, Wang E, Goldberg AR et al. Changes in the amino terminal sequences of $pp60^{src}$ lead to decreased membrane association and decreased in vivo tumorigenicity. Cell 1982; 889-896.
70. Cross FR, Garber EA, Hanafusa H. N-ter-

minal deletions in Rous sarcoma virus pp60^{src}: Effects on tyrosine kinase and biological activities and on recombination in tissue culture with the cellular *src* gene. Mol Cell Biol 1985; 5: 2789-2795.

71. Jove R, Garber EA, Iba H et al. Biochemical properties of pp60^{src} mutants that induce different cell transformation parameters. J Virol 1986; 60: 849-857.
72. Jove R, Mayer BJ, Iba H et al. Genetic analysis of pp60$^{v\text{-}src}$ domains involved in the induction of different cell transformation parameters. J Virol 1986; 60:840-848.
73. Fukui Y, O'Brien MC, Hanafusa H. Deletions in the SH2 domain of p60$^{v\text{-}src}$ prevent association with the detergent-insoluble cellular matrix. Mol Cell Biol 1991; 11: 1207-1213.
74. Hamaguchi M, Hanafusa H. Association of pp60^{src} with triton X-100-resistant cellular structure correlates with morphological transformation. Proc Natl Acad Sci USA 1987; 84: 2312-2316.
75. Hamaguchi M, Hanafusa H. Localization of major potential substrates of p60$^{v\text{-}src}$ kinase in the plasma membrane matrix fraction. Oncogene Res 1989; 4: 29-37.
76. Resh MD. Specific and saturable binding of pp60$^{v\text{-}src}$ to plasma membranes: evidence for a myristyl *src* receptor. Cell 1989; 58: 281-286.
77. Resh MD, Ling H-P. Identification of 32K a plasma membrane protein that binds to the myristylated amino-terminal sequence of pp60$^{v\text{-}src}$. Nature (London) 1990; 346: 84-86.
78. Sigal CT, Resh MD. the ADP/ATP carrier is the 32 kilodalton receptor for an NH2-terminally myristylated *src* peptide but not for pp60^{src} polypeptide. Mol Cell Biol 1993; 13: 3084-3092.
79. Warren SL, Nelson WJ. Nonmitogenic morphoregulatory action of pp60$^{v\text{-}src}$ on multicellular epithelial structures. Mol Cell Biol 1987; 7: 1326-1337.
80. Warren SL, Handel LM, Nelson WJ. Elevated levels of pp60$^{c\text{-}src}$ alter a selective morphogenetic property of epithelial cells in vitro without a mitogenic event. Mol Cell Biol 1988; 8: 632-646.
81. Falcone G, Tato F, Alema S. Distinctive effects of the viral oncogenes *myc*, *erb*, *fps* and *src* on the differentiation program of quail myogenic cells. Proc Natl Acad Sci USA 1985; 82: 426-430.
82. Yoshimura M. Change of hyaluronic acid synthesis during differentiation of myogenic cells and its relation to transformation by Rous sarcoma virus. Cell Differentiation 1985 16: 175-185.
83. Alema S, Casalbore P, Agostine E et al. Differentiation of PC12 pheochromocytoma cells induced by *v-src* oncogene. Nature (London) 1985; 316: 557-559.
84. Trotter J, Boulter CA, Southeimer H et al. Expression of *v-src* arrests murine glial cell differentiation. Oncogene 1989; 4: 457-464.
85. Carmier JF, Samarut J. Chicken myeloid stem cells infected by retroviruses carrying the *v-fps* oncogene do not require exogenous growth factors to differentiate in vitro. Cell 1986; 44: 159-165.
86. Menko AS, Boettiger D. Inhibition of chicken embryo lens cell differentiation and lens junction formation in culture by pp60$^{v\text{-}src}$. Mol Cell Biol 1988; 8: 1414-1420.
87. Atkinson MM, Menko AS, Johnson RG et al. Rapid and reversible reduction of junctional permeability in cells infected with a temeprature-sensitive mutant of avian sarcoma virus. J Cell Biol 1981; 91: 573-578.
88. Weber MJ, Evans PK, Johnson MA et al. Transport of potassium, amino acids and glucose in cells transformed by Rous sarcoma virus. Federation Proc. 1984; 43: 107-112.
89. Spivack JG, Maller JL. Phosphorylation and protein synthetic events in Xenopus laevis oocytes microinjected with pp60$^{v\text{-}src}$. Mol Cell Biol 1985; 5: 3629-3633.
90. Balmain A, Ramsden H, Borden GT et al. Activation of the mouse cellular Harvey *ras* gene in chemically induced benign skin papillomas. Nature (London) 1984; 307: 658-661.
91. Land H, Parada LF, Weinberg RA. Tumorigenic conversion of primary embryonic fibroblasts require at least two cooperating oncogenes. Nature (London) 1983; 304: 596-602.
92. Land H, Parada LF, Weinberg RA. Cellular oncogenes and multistep carcinogenesis.

Science 1983; 222: 771-777.
93. Ruley HE. Adenovirus early region 1A enables viral and cellular transforming genes to transform primary cells in culture. Nature 1983; 304: 602-606.
94. Graf T. Leukemia as a multisetep process: studies with avian retroviruses containing two oncogenes. Leukemia 1988; 2: 127-132.
95. Hunter T. Cooperation between oncogenes. Cell 1991; 64: 249-270.
96. Dolberg DS, Bissell MJ. Inability of Rous sarcoma virus to cause sarcomas in the avian embryo. Nature (London) 1984; 309: 553-556.
97. Dolberg DS, Hollingsworth R, Hertle M et al. Wounding and its role in RSV-mediated tumor formation. Science 1985; 230: 676-678.
98. Sieweke M, Stoker AW, Bissell MJ. The cocarcinogenic effect of wounding and RSV-tumorigenesis: potential mechanisms. Cancer Res 1989; 49: 6419-6424.
99. Stoker AW, Hatier C, Bissell MJ. The embryonic environment strongly attenuates *v-src* oncogenesis in mesenchymal and epithelial tissues but not in endothelia. J Cell Biol 1990; 111: 217-228.
100. Soriano P, Montgomery C, Geske R et al. Targeted disruption of the *c-src* proto-oncogene leads to osteopetrosis in mice. Cell 1991; 64: 693-702.
101. Grant SGR, O'Dell TJ, Stein PL et al. Impaired long-term potentiation, spatial learning, and hippocampal development in *fyn* mutant mice. Science 1992; 258: 1903-1910.

CHAPTER 3

Structural Changes in RSV-Transformed Cells

MORPHOLOGICAL CHANGES IN RSV-TRANSFORMED CELLS

When cells are transformed by oncogenic viruses a variety of changes occur which result in a characteristic phenotype. These include induction of tumor formation in vivo, changes in the regulation of cellular proliferation in vitro, loss of density-dependent inhibition of growth, increased metabolite uptake, a reduction in growth factor requirement, changes in surface glycoprotein expression, increases in membrane fluidity and release of protease activity. The most obvious changes when cells are transformed by RSV are those of morphology, motility and adhesion (for reviews see references 1,2,3), all of which may in some way be linked to unregulated cell proliferation. Transformed cells acquire the ability to grow in semi-solid media such as methyl cellulose or soft agar.[4,5] Although nontransformed cells can grow under soft agar they reach a confluent density and stop growing. Transformed cells, however, continue to proliferate due to the loss of dependency on the tissue culture substratum for proliferation. This can be seen in Figure 3.1 which shows a colony of RSV-transformed cells growing up into an overlay of soft agar. Soft agar growth also demonstrates that the cells also have an altered morphology, being rounded up compared with nontransformed cells. This may be related to the increased invasiveness of RSV-transformed cells both in vitro[6] and in vivo.[7]

Several investigators have reported that cell shape and control of macromolecular metabolism are closely linked. In a study of normal 3T3 mouse fibroblasts O'Neill et al[7] found that precisely controlling cell shape by limiting the availability of adhesive substrate resulted in tight regulation of DNA synthesis. Anchorage-induced stimulation of proliferation occurred when the available substratum was greater than or equal to 500 μm^2, thus normal cells have a finite requirement for adhesion before proliferation can occur. As I will describe later, this is almost certainly related to the adhesive contacts between cells and the tissue culture substratum, and particularly reorganization of cytoskeletal components which mediate this adhesion. Others have shown that neoplastic transformation results in an uncoupling between cell shape and proliferation.[9-11] In ad-

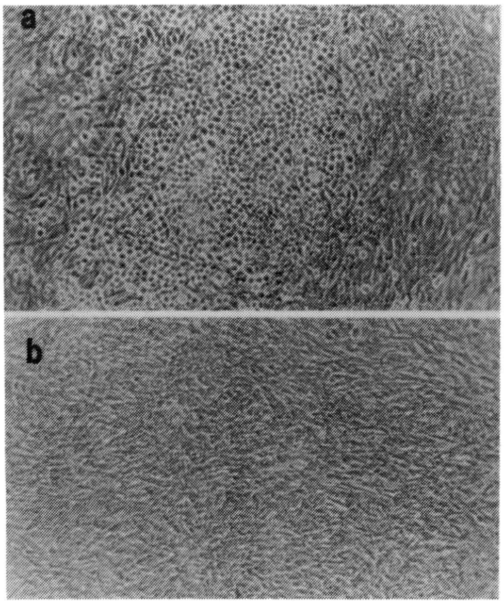

Fig. 3.1. Growth of cells under agarose. a: A colony of RSV-transformed cells showing overgrowth up into the semi-solid medium and rounded morphology. b: Nontransformed cells grow as a monolayer.

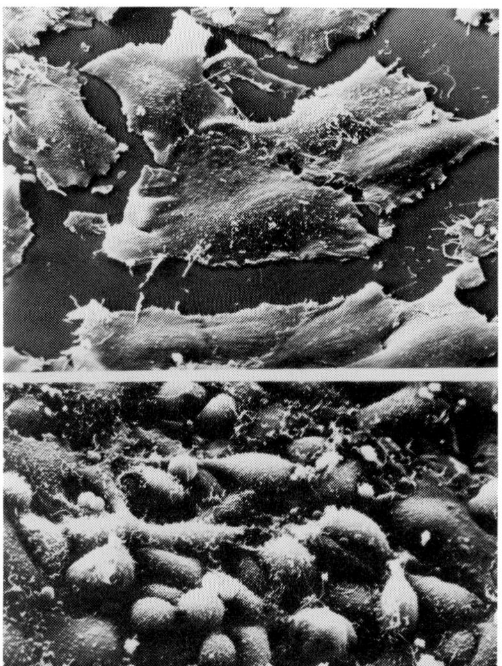

Fig. 3.2. Scanning electron micrograph of nontransformed cells (upper panel) and RSV-transformed cells (lower panel) showing rounded morphology and increased membrane activity after transformation. Figure courtesy of Drs G Felice and M Nermut; S. Kellie. Treatise on the cytoskeleton. Vol. 4, JAI Press.

dition to rounding up, other structural changes which occur are alterations in contact inhibition of movement, leading to piling up of cells in confluent cultures,[3] and increased surface membrane activity in the form of ruffles and pseudopodia. These alterations are easily shown using scanning electron microscopy as demonstrated in Figure 3.2. The morphological changes accompanying transformation are due to alterations in both surface extracellular matrix components and the intracellular organization of the cytoskeleton.

EXTRACELLULAR MATRIX COMPONENTS AND TRANSFORMATION

The reduction in adhesiveness of transformed cells has been attributed to the observations that these cells commonly display a reduction in extracellular matrix (ECM) components, the most obvious being a loss of fibronectin from the cell surface.[12,13] Immunofluorescence of fibronectin in nontransformed fibroblasts or RSV-transformed fibroblasts demonstrates a dramatic loss of fibronectin matrix after transformation (Fig. 3.3). This is due to both a decrease in biosynthesis and an increase in degradation of the protein. The addition of exogenous fibronectin to cultures of RSV-transformed cells restores a morphology similar to though not identical with that of normal cells, including increased stress fiber formation and focal contacts, indicating that fibronectin loss makes some contribution to the transformed phenotype (Fig. 3.4).[14-19] Transformed cells commonly express surface proteases at or near membrane-matrix interactions, and these may be responsible for the reduction in fibronectin from transformed cells. There is some evidence that in addition to surface proteases which are endogenously synthesized by normal cells, novel proteases are expressed by transformed cells, and it is of interest to note that normal proteases induced in RSV-transformed cells are expressed at sites of adhesion, where these enzymes would be most likely to hydrolyse fibronectin or other ECM components.[20-23] Experiments which demonstrate that treatment of normal fibroblasts with proteases leads to responses similar to transformation including increased proliferation[24,25] and cytoskeletal alteration support the suggestion that protease activity is related to transformation,[26] and other studies showing that the addition of protease inhibitors to cells reduced proliferation and increased cell adhesiveness.[27-31] Protease activity is also supported by our findings that a single dose of fibronectin is not sufficient for morphological reversion of transformed cells to a more normal phenotype, but this ECM component needs to be added regularly over a period of many hours or days, consistent with increased proteolytic enzymes on the transformed cells continually degrading the exogenously added material (S. Kellie, unpublished).

The protease plasmin is associated with the extracellular matrix and its role in transformed cells has been studied by a number of groups.[32,33] Plasmin is normally inactive but can be activated by another protease, plasminogen activator,

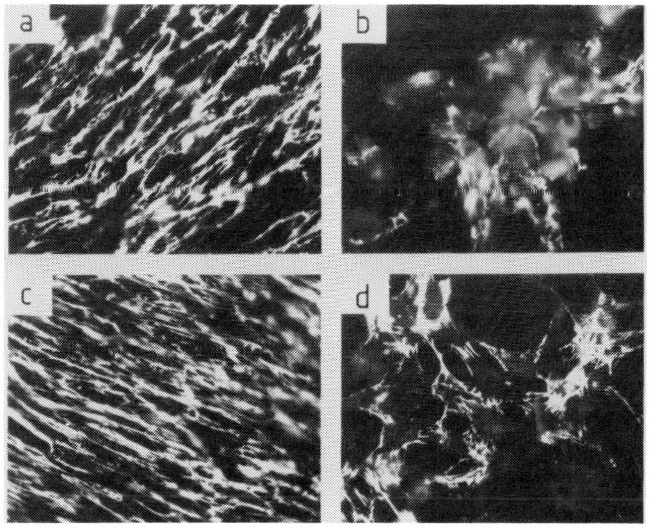

Fig. 3.3. Reduction in fibronectin matrix production after transformation by RSV. Chick embryo fibroblasts were transformed by wild-type RSV and confluent cultures of nontransformed cells (a) or RSV-transformed cells (b) were stained for fibronectin. Panels c and d show fibronectin staining of confluent cultures of cells transformed by the temperature-sensitive mutant LA29 and cultured at the restrictive temperature (c) or permissive temperature (d) for transformation.

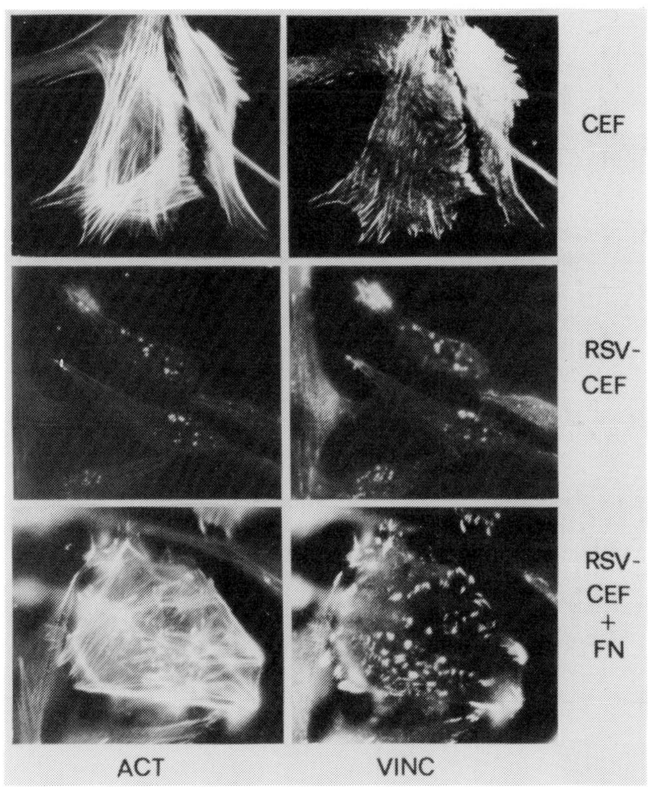

Fig. 3.4. RSV-transformed cells after addition of exogenous fibronectin. F-actin staining (left hand panels) and vinculin localization (right hand panels) were visualized in the same cells by double fluoresence labelling. The upper panels show nontransformed cells cultured without exogenous fibronectin. The middle panels show RSV-transformed cells cultured without exogenous fibronectin. The bottom panels show the partial restoration of stress fibers and vinculin-containing adhesion plaques in RSV-transformed cells cultured in the presence of exogenous fibronectin. Fig. courtesy of Royal Society of Chemistry.

which can be elevated after transformation. Recently these enzymes have gained more prominence due to the finding that urokinase, a member of the plasminogen activator family, is found at sites of cell-substratum adhesion.[34-38] These experiments all strongly suggest that proteases at the cell surface and in particular at the areas where cells attach to the extracellular matrix are activated in transformed cells. This activity may well be relevant to the ability of transformed cells to invade surrounding tissues, including breaking down basement membranes and also may be relevant to the ability of transformed cells to metastasize to secondary sites.

THE CYTOSKELETON OF TRANSFORMED CELLS

Early studies used transmission electron microscopy to reveal that there were changes in the fine structure of the cytoskeletons of transformed cells compared to normal counterparts. McNutt et al[39,40] found that in particular there were decreases in the expression of the microfilament sheaths which are distributed under the plasma membrane as a cortical network of filamentous actin. Wang and Goldberg[41] also found a reduction in microfilament bundles running under the plasma membrane, and very few bundles in the cytoplasm in cells transformed by other oncogenic viruses. In addition they showed that there was an increased number of microfilaments which were arranged in disorganized meshworks in the cortical region of the transformed cells. Improved techniques in electron microscopy have allowed cellular ultrastructure to be visualized more clearly, and to investigate the fine structure of the cytoskeleton in more detail, Felice et al[42] examined the filamentous system of transformed cells by using electron microscopy to study quick frozen whole cytoskeletons and showed that transformation is accompanied by a change in microfilaments from well-organized bundles containing up to 10 filaments to an extensive, convoluted meshwork of mainly single filaments containing multiple apparently random interconnections.

Thus the filamentous system which is most obviously changed upon transformation is the microfilament system. This system is a contractile network based on actin and other associated proteins which is still tightly regulated in a cell-type or cell cycle-dependent manner. To understand more clearly the transformation-associated changes it is worthwhile briefly to review the role and composition of microfilaments in normal cells before examining the changes induced by oncogenic tyrosine kinases.

Actin and Microfilament Bundles

Actin is often the most abundant protein in non-muscle cells, and can exist either as a 45kDa monomer (G actin) or in a polymeric form (F actin). In contrast to the isotropic filamentous arrangement of actin in striated muscle cells, the spatial organization of actin in most nonmuscle cells such as fibroblasts is highly variable and may serve both structural and contractile functions. These actin isoforms have similar polymerization properties in vitro and bind heavy meromyosin subfragment 1 which can be seen as arrowhead structures when visualized by electron microscopy, giving an indication of the polarity of filaments within cells. In most nonmuscle cells, actin filaments are dynamic, contractile and undergo constant assembly and disassembly with rapid turnover.[43] Many aspects of cell function such as cell shape, motility, phagocytosis and cell-matrix interactions, i.e., many of the functions which change after oncogenic transformation, are dependent on the precise regulation of these actin filament networks.

The extent of actin polymerization appears to be an important and regulated aspect of cell function. In the cytoplasm polymerized actin exists in equilibrium with the monomeric form, but the equilibrium constant for the polymerization-depolymerization reaction can be regulated by proteins that bind selectively to monomer or polymer. Depending on cell type, approximately half of the actin in cells is unpolymerized although the intracellular ionic conditions and nucleotide concentrations would predict the preferential formation of polymeric actin. This occurrence of more monomeric actin than predicted in the cytoplasm is due to the presence of proteins which bind to G-actin resulting in only partial polymerization. For example profilin prolongs the lag phase of actin polymerization and also inhibits the extent of polymerization. Profilin also binds certain lipids which regulate its interaction with actin and the importance of this to transformation will be discussed in the next chapter.[44] Several actin-binding proteins control the polymerization of actin including profilins, DNaseI, Thymosin β4 and vitamin-D-binding protein. Proteins such as gelsolin and villin sever long actin filaments and prevent their reassembly at high Ca^{2+} concentrations. However at low Ca^{2+} concentrations they can also act as nucleating centers, accelerating the polymerization of monomeric actin into filaments. Proteins such as filamin and α-actinin crosslink filamentous actin into microfilament bundles and these may also interact with a group of proteins including vinculin and talin that are presumed to be involved in the attachment of actin to membranes.

In addition to changes in the actin monomer/polymer ratio the organization of actin within cells can also be regulated at the level of bundling and cross-linking of single actin filaments into multifilamentous networks, an example of this being microfilament bundles in fibroblasts. These structures are composed of bundles of actin filaments originally termed stress fibers, which often extend the length of the cell[45,46] and can be visualized by phase contrast microscopy but are better seen by immunofluorescence using antibodies against actin or the F-actin-specific probe phalloidin. Microfilament bundles are believed to be involved in cell-substratum and intercellular adhesions since they terminate at plasma membrane specializations structures where these adhesions occur. The preferred end for actin polymerization is closest to the plasma membrane[47] and electron microscopy has shown that all microfilament bundles have the same orientation towards the membrane. Immuno-fluorescence studies show that stress fibers also contain other muscle-related proteins such as myosin,[48] α-actinin,[49] filamin[50] and tropomyosin[51] and these proteins appear to have a periodicity in their association with the filament bundles. Stress fibers are probably more important for cell adhesion than for motility since actively motile cells have few stress fibers while non-motile fibroblasts tend to have many.[52,53] The introduction of signaling molecules into cells has a profound effect on the stress fiber network and this will be described later in this chapter.

Actin Attachment to the Plasma Membrane

How do actin filaments associate with or attach to the plasma membrane? In specialized cells such as the intestinal brush border proteins such as myosin I[54,55] have been implicated, but the membrane components with which this protein interacts have not yet been identified. In erythrocytes actin appears to associate with the membrane by interacting with a membrane complex consisting of the transmembrane anion exchanger, spectrin, ankyrin and band 4.2 or by associating with a complex of spectrin, protein 4.1 and glycophorin C or the anion exchanger.[56] Other cells, particularly those cells which are adherent in tissue culture, have a different set of membrane anchoring proteins. In fibroblasts stress fibers terminate in membrane specializations termed focal adhesions or adhesion plaques. These are typically elongated or arrowhead-shaped areas where the plasma membrane is in close apposition (<15 nm) to the tissue culture substratum. Adhesion plaques also contain concentrations of a number of cytoskeletal proteins which are predicted to mediate the association of actin filaments with the plasma membrane.

The list of proteins found in the adhesion plaque is increasing rapidly, and examples of proteins localized on the cytoplasmic face, proteins localized on the extracellular surface, and transmembrane proteins have all been described. The best-studied adhesion plaque proteins are α-actinin, vinculin, talin, paxillin and the transmembrane receptors for extracellular matrix components (integrins). A hypothetical model for the interactions of these proteins within an adhesion plaque is shown in Figure 3.5, where actin microfilaments bind to a complex of proteins at the plasma membrane. Examples of putative intermolecular associations which occur in this membrane specialization are the binding of paxillin to the rod domain of vinculin[57] and the interaction of α-actinin with the integrin β-subunit, and these associations might mediate actin-plasma membrane interactions within cells.[58] The coated pit-associated protein clathrin is also consistently found at cell-substratum adhesions, although its function in this location is unknown.[59] Several groups have described different forms of integrin which have actin tightly bound suggesting a direct attachment of actin to matrix receptors in a manner independent of vinculin or talin, at least in cells such as platelets.[60,61] However it is more likely that the interaction of actin with the adhesion plaque occurs according to the model shown in Figure 3.5 with the actin associating with the fibronectin receptor via talin, vinculin and possibly α-actinin intermediaries.

Nicol and Nermut[59] have described another adhesion structure in preparations of adherent plasma membranes. In addition to typical elongate focal adhesions containing actin and vinculin, adherent membrane patches lacking microfilament bundles and negative for vinculin labeling were found. These were coated with clathrin and identifiable in interference reflection microscopy as less dark than focal adhesions and having dot, U- or sinusoidal shapes. There is no information whether these novel structures have a different adhesive function.

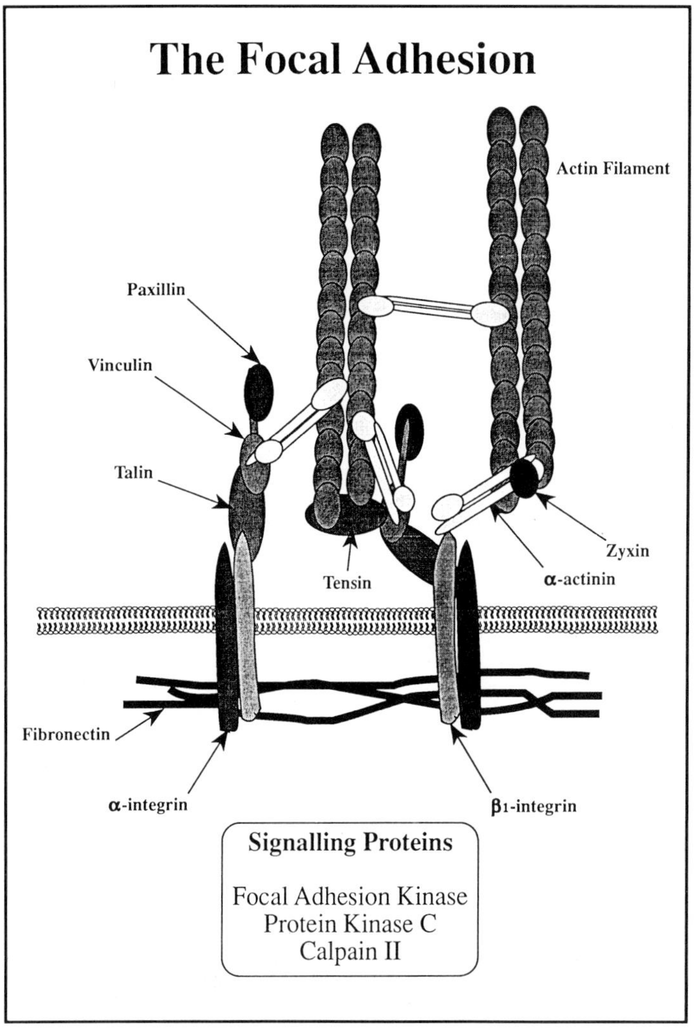

Fig. 3.5. Model of molecular interactions which occur in focal contacts. In addition to the proteins indicated a number of signaling proteins such as focal adhesion kinase, protein kinase C and calpain II have been localized to adhesion plaques. Figure courtesy of S. Barry and D .R. Critchley, University of Leicester, UK.

Proteins Involved in Cytoskeleton-Membrane-Matrix Interactions

α-actinin

α-actinin from both muscle and non-muscle sources is a homodimer of 100 kDa subunits and has the ability to crosslink actin filaments into bundles in vitro. α-actinin was originally described as a major protein of the Z band of striated muscle, which is the site of apparent attachment of microfilaments to the sarcolemma in these cells,[62] and this localization indicated that it might have a similar function in non-muscle cells. Immunofluorescence studies showed that α-actinin distributes periodically along stress fibers and there is an apparent increase in staining intensity where these bundles terminate (Fig. 3.6).[51] α-actinin can be detected in membrane preparations derived from several different cell types, including cultured Hela cells, lymphocytes, and fibroblasts[63] and an interaction between α-actinin and the β_1 integrin subunit has been shown in vitro,[58] both of which suggest a direct role for α-actinin in the attachment of microfilaments to the membrane. α-actinin has structural and amino acid homologies with spectrin and dystrophin which also closely associate with the plasma membrane, and so there may be conserved domains of proteins which mediate association with the plasma membrane.[64,65]

Vinculin

Vinculin is a 116 kDa protein present in cell-matrix adhesions and at cell-cell junctions,[66,67] and this localization has led to suggestions that it plays a role in the anchoring of microfilaments to the membrane. Two-dimensional SDSPAGE has revealed three isoforms of vinculin; α, β and γ, which have identical molecular weights, and appear to differ in the degree of phosphorylation and in their association with either the cell membrane or cytoskeleton.[68] Over 90% of the total vinculin within cells is detergent soluble and this comprises mainly the β form, whereas the cytoskeleton-associated vinculin contains all three isoforms of vinculin. It is not known whether this apparent differential phosphorylation is responsible for differential cytoskeleton association. A 159 kDa vinculin-related protein has been identified in smooth muscle and platelets. This protein, termed metavinculin, is recognized by monoclonal antibodies to vinculin and has a similar proteolytic peptide profile to vinculin. More recently metavinculin has been shown to be generated by alternative splicing of the vinculin gene.[69,70] Less is known about the function of metavinculin than vinculin.

Early experiments suggested that vinculin could bind to actin and induce its bundling[71] or capping.[72] Later studies, however, showed that these activities were not due to vinculin itself but rather to a low molecular weight contaminant that copurified with vinculin.[73,74,72] There have been reports of a direct interaction between vinculin and actin,[75,76] however arguments against a direct association of vinculin with

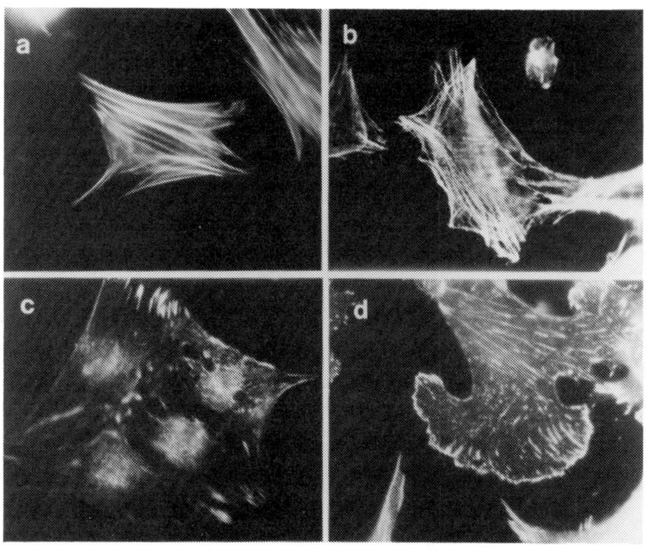

Fig. 3.6. Immunofluorescent localization of actin (a), α-actinin (b), vinculin (c) and talin (d) in chick embryo fibroblasts.

actin are supported by experiments showing that both fragmin and microfilament fragmenting protein can be used to remove actin from adhesion plaques without the loss of vinculin.[77] Furthermore, vinculin is readily solubilized, whereas many of the cytoskeletal proteins, including actin remain insoluble.[78] An interaction between vinculin and actin has been suggested by the findings that when epithelial cells are incubated with EGTA to disrupt intercellular junctions, microfilaments associated with those junctions are released from the cytoplasmic side of the membrane in association with vinculin.[79] Antibody mapping of functional domains in vinculin suggest that residues 587-851 could be an actin-binding site[80] and this site contains a putative phosphorylation site on tyrosine 822.[81] This may well be the site for phosphorylation by pp60$^{v\text{-}src}$, however as will be described in the next chapter the functional significance of this phosphorylation is questionable.

Vinculin is found at the ends of microfilament bundles, and coaligns with cell surface fibronectin fibrils[82] which in turn co-distribute with underlying actin filaments.[83-85] This location of vinculin suggests that even though it may not bind actin directly, it does have a function in the attachment of bundles of actin filaments to the cytoplasmic face of the plasma membrane in these locations. Consistent with this hypothesis are the observations that vinculin is absent from motile lamellipodia and shows no preferential association with precursors of focal contacts which have no associated microfilament bundles, but does accumulate within new contacts as they mature and as multifilamentous complexes of actin appear.[86] Whilst vinculin is not an integral membrane protein, in vitro studies have shown that it can insert into a lipid bilayer by interacting with acidic phospholipids, and it is possible that this is mediated by other lipids such as palmitic acid or myristic acid bound covalently to vinculin.[87-89] The exact mechanism by which vinculin associates with focal contacts is unknown, however the NH2-terminal head domain of vinculin can bind to talin[75,90,91] and this interaction might result in the retention of vinculin within adhesion plaques. Vinculin molecules can also form oligomeric aggregates held together by sequences near the carboxy terminus and so it is possible that vinculin might also undergo self-association leading to positively cooperative binding within focal contacts.[92]

Talin

Talin is a 215 kDa protein with some homology to erythrocyte band 4.1 and ezrin and which is concentrated in adhesion plaques and ruffling membranes of cells (Fig. 3.6).[93] Talin binds to vinculin as described in the previous section and has also been shown to bind to the chicken fibronectin receptor.[90,94] Talin has also been shown to bind to actin in vitro indicating that it may directly organize actin filaments at sites of membrane interaction.[95] Proteolytic digestion of talin results in a 200 kDa and a 47 kDa fragment which are generated from two distinct domains of the protein. The 200 kDa fragment binds to both vinculin and metavinculin in vitro and this polypeptide is incorporated into focal adhesions when microinjected into fibroblasts. The smaller fragment does not bind to vinculin but retains the ability to associate with focal adhesions of fibroblasts.[96] These experiments indicate that there are at least two sites on the talin molecule which contribute focal contact localization. Talin dissociates from adhesion plaques more slowly than vinculin after RSV-induced transformation, and adhesion plaques with attached microfilaments can exist with very little or no vinculin present suggesting that talin or another protein is responsible for actin-adhesion plaque association[97]. However, at present most of the evidence supports the role of talin as shown in Figure 3.5.

Fibronectin receptor

Several integral membrane proteins have been identified in focal adhesions, and the most obvious of these belong to the integrin family of extracellular matrix receptors.[98] The first focal adhesion integrin to be demonstrated was the fibronectin receptor which, like all integrins, is a heterodimer comprising α and β subunits associating into a complex of a globular head with two tails which contain the transmembrane portion of the molecule.[99] Two main fibronectin receptors have now been established: high affinity α5β1 integrins and lower affinity α3β1 integrins, although both bind to an RGDS-containing sequence in the cell adhesive domain of fibronectin.[100,101] Immunofluorescence experiments have shown that fibronectin and its receptor are co-distributed in fibrillar arrays in the focal contacts of cultured fibroblasts.[102] Rapidly migrating embryonic cells, as well as RSV-transformed fibroblasts, display a diffuse organization of the receptor in vitro that contrasts with its clustering in linear arrays in putative transmembrane linkage complexes on normal fibroblasts, suggesting that the cell surface distribution of this receptor is related to cell behavior. A similar diffuse pattern of cell surface localization was found in vitro on RSV tumor cells and in transformed cultured human cells.[103,104]

The fibronectin receptor also contributes to matrix assembly,[105] and can be responsible for outside-in in addition to inside-out signaling, perhaps due to some cytoskeletal interaction.[106-108] Synthetic GRGDS peptides which block the binding of integrin receptors to extracellular matrix components induce dissociation of α-actinin and vinculin from the sites of focal contacts.[109] This suggests that the binding state of integrin has a transmembrane effect on the distribution of cytoskeletal components. This dissociation of α-actinin and vinculin from adhesion plaques may in turn weaken the contacts and result in rounding and detachment of cells. Fibulin, a 100 kDa protein, also interacts with the cytoplasmic domain of the integrin β subunit and so may be an additional mediator of interactions between adhesion receptors and the cytoskeleton.[110]

Fibronectin

The extracellular matrix component fibronectin is a dimer of two polypeptides disulphide bonded near the carboxy terminus to generate an overall size of 550 kDa.[11] There are two main sites in fibronectin that promote cell attachment, one of which contains the sequence Arg-Gly-Asp (RGD), a recognition motif for many integrins resulting in the attachment of cells to the extracellular matrix.[101,106] Fibronectin synthesized by cells tends to distribute on the cell surface such that it co-aligns with stress fibers or adhesion plaques.[82-85]

As discussed earlier, in addition to reduced synthesis, degradation of the matrix components deposited at the cell surface probably contributes to the lack of surface-associated matrix in neoplastic cells, since adhesion plaques contain the plasminogen activator urokinase,[35] and other cell-surface proteases capable of degrading fibronectin.[21]

Other Proteins

Several other proteins have been localized to the focal contact and some examples are listed here. A recent review provides an extensive review of membrane-cytoskeletal associations (see reference 50 in chapter 4). Zyxin, an 82 kDa protein binds to the amino terminus of α-actinin and also to actin. Tensin is a 150/200 kDa protein which associates with the barbed end of actin filaments and has been shown to bind to vinculin. This molecule also contains an SH2 domain and so has the potential to associate with tyrosine-phosphorylated proteins, such as focal adhesion kinase and *fyn* which are found in focal contacts. Some protein kinases such as protein kinase C, focal adhesion kinase and pp60$^{v\text{-}src}$ are concentrated in cellular adhesion plaques and the relevance of this will be discussed later.

CHANGES IN MICROFILAMENT BUNDLES AND ACTIN FILAMENTS

Although changes in a number of filamentous structures including microtubules and intermediate filaments have been re-

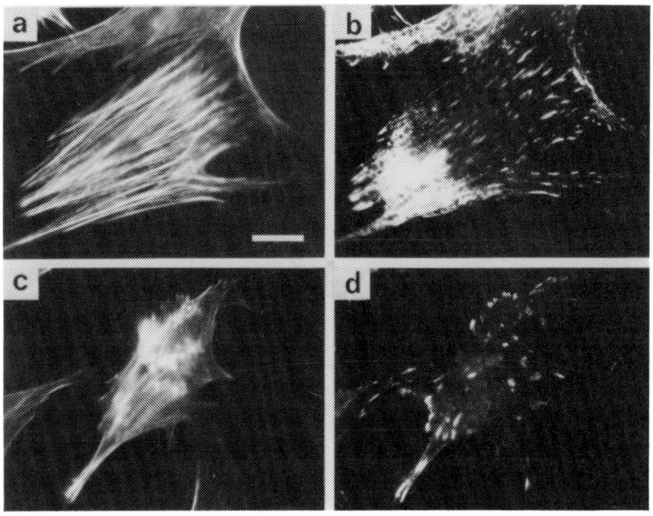

Fig. 3.7. Double labelling of actin (a,c) and vinculin (b,d) in nontransformed cells (upper panels) or RSV-transformed cells (lower panels). Figure courtesy of J Cell Science and the Company of Biologists.

ported in transformed cells, changes in the actin-based cytoskeleton have been the most extensively studied, and I will focus on changes in actin and its associated proteins, particularly the components which are involved in actin-plasma membrane association in this monograph. In addition to a rounded morphology as discussed earlier, transformed cells commonly exhibit cytoskeletal changes including a reduction in microfilament bundles and the redistribution of actin-associated proteins such as α-actinin, vinculin and talin (Fig. 3.7). One of the most consistent cytoskeletal markers for cell transformation in vitro appears to be the disappearance or diminution of actin-containing stress fibers. Once antibodies against cytoskeletal proteins had been developed several groups investigated the organization of the proteins of the filamentous systems after transformation by a variety of agents. One of the first studies to be performed used a mouse 3T3 cell line that had been transformed by SV40 virus. Although this does not encode an oncogenic tyrosine kinase, these early studies are relevant to other transforming viruses such RSV. Pollack et al[111] found that nontransformed cells showed phase-dense sheaths and stress fibers which stained strongly with antibodies against actin and myosin and measured about 1 µm by 50 µm. SV40-transformed derivatives, however had virtually no actin-containing stress fibers and exhibited a diffusely staining actin-containing matrix. These workers concluded that this was due to a redistribution of actin rather than a reduction, since both the nontransformed and transformed cells contained the same total amount of actin and myosin. In addition these workers surveyed a variety of cell characteristics including anchorage-dependent growth, density-dependent inhibition of proliferation and serum requirements and concluded that only anchorage-dependent growth correlated with loss of actin-containing microfilament bundles.

In studies of RSV-transformation, mutants have been useful in examining the transformation-dependent changes in the actin cytoskeleton. In particular temperature-sensitive *(ts)* mutants have allowed us to investigate the early stages of transformation. In cells infected with *ts* RSV the transformed phenotype is expressed when cells are cultured at the permissive temperature of 35°C, but they are phenotypically normal at the restrictive temperature of 39-41°C. Most studies have indicated that this is due to the production of a thermolabile *src* protein which is enzymatically active at the permissive temperature but is inactive at the restrictive temperature, however there are some exceptions to this where the lesion may be at the level of intracellular regulation or substrate recognition rather than enzyme activity.[112] Thus by culturing cells at the restrictive temperature then switching to the permissive temperature characteristics of transformation can be studied within the same cell line. Using such a system Ambros et al[113] showed changes in plasma membrane conformation characterized by evaginations after the onset of transformation.

Wang and Goldberg[41] studied *ts*-transformed chick embryo fibroblasts and defined three stages in architectural changes when cells were switched from restrictive to permissive temperature. Stage 1, less than one hour after temperature shift: flower-like membrane ruffles in the nuclear region. During this stage actin-containing stress fibers were maintained and normal-looking. Stage 2, 3-12 hours after temperature shift: Appearance of microvilli, retraction of the cell body and gradual loss of fibroblastic cell shape. The microfilament bundles gradually shortened and the number within cells was reduced. Stage 3, 12-24 hours after temperature shift: Rounding up of the cells and appearance of small blebs. Virtually no microfilament bundles were present at this time with only diffuse actin staining in the cell cytoplasm. Thus there was a progressive reduction in microfilament bundles over a 24-hour period after the initiation of transformation. These experiments indicated that the *src* gene product, either directly or indirectly, could alter both the assembly and organization of microfilament bundles.[114]

In a similar study of the early events induced by transformation using *ts* mutants of RSV Boschek et al[115] observed that within 15 minutes after the initiation of transformation ruffle-like "whorls" appeared on the cell surface. The whorls gradually increased up to three hours, when about 60% of the cells contained them, then decreased over the next 24 hours. The whorls stained intensely for actin. Moreover, fluorescent phalloidin (which binds specifically to F-actin) showed that these structures were composed of filamentous actin and in addition they also contained α-actinin, myosin and tropomyosin. Another early event in neoplastic transformation is the induction of leading edge membrane ruffles which were maximal at 15 minutes then decreased. Although these ruffles contained actin, other proteins such as α-actinin, myosin and tropomyosin were not present. At later periods (6-12 hours after switch to permissive temperature) the actin cables became disarranged with a breakdown in the periodic distribution of α-actinin, myosin and tropomyosin and a concurrent increase in diffuse cytoplasmic staining. At these later stages there was an increase in small dots of actin which may be remnant focal contacts or podosomes as will be discussed later. The intensity of staining with fluorescent phalloidin decreased after the induction of transformation indicating that there was a shift in the equilibrium from F- to G-actin. Others have obtained similar results using different *ts* mutants and cells. It has been suggested that this later stage is a consequence of depolymerization of filamentous actin, despite the existence of some filaments in transformed cells;[41,116,117,118] however, there is little evidence to support this. To investigate the mechanism of microfilament bundle loss in RSV-transformed cells we compared the functional state of actin in transformed cells and reported that there were differences in the actin monomer: polymer equilibrium when an non-transformed cell line was compared with an RSV-transformed derivative, but that the changes were complex and not simply a net depolymerization.[42] This was investigated in more detail by examining the cytoskeletons of a cell line transformed by a *ts* mutant of *v-src*, *ts*LA29 which has properties similar to the *ts*RSV already described. At the restrictive temperature the actin bundling protein α-actinin decorated the detergent-insoluble microfilament bundles in a periodic pattern, however at the permissive temperature α-actinin was found predominantly at the cell periphery, although occasionally brightly staining aggregates were present within the cells which co-localized with actin and also vinculin-containing focal contacts. At 39°C many well developed vinculin-containing adhesion plaques were present, mainly located at α-actinin-containing termini and also in cell-cell contacts. At 35°C there was a substantial reduction in the number of adhesion plaques and a change in their appearance from elongated arrowhead-like structures to smaller dot-like aggregates which were located almost exclusively at the cell periphery, similar to those seen in the transformed cells depicted in Figure 3.7. Thus at the restrictive and permissive temperatures the cells had the characteristics of nontransformed and transformed cells respectively. The distribution and three-dimensional arrangement of microfilaments and other filamentous systems in the cytoskeletons of these cells was also studied using whole-mount electron microscopy. Cells grown at 39°C showed a typical image of long multifilamentous bundles running from the nucleus towards the cell periphery, analogous to those seen by fluorescence. These bundles were also prominent at the cell periphery running under the solubilized plasma membrane (Fig. 3.8). These were microfilaments since they could be labeled with heavy meromyosin subfragment 1. Single actin filaments were rare but many single intermediate filaments were observed. At the permissive temperature the stress fibers in these cells had virtually disappeared and when present were much thinner with aggregated filamentous material more apparent. Single actin filaments in these aggregates could be identified by heavy meromyosin subfragment 1 labeling.

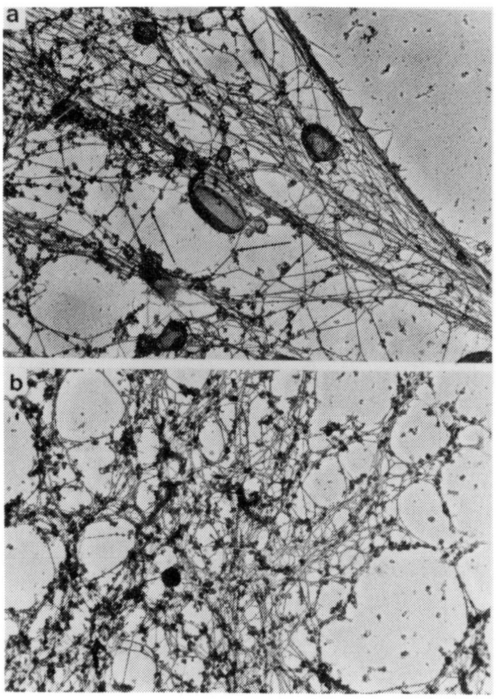

Fig. 3.8. Whole-mount electron microscopy of the cytoskeletons from nontransformed (a) or v-src-transformed rat fibroblasts (b). Nontransformed cytoskeletons are organized into bundles of microfilaments whereas cytoskeletons from transformed cells are revealed as a convoluted network with apparently random interconnections. Figure courtesy of Drs. G. Felice and M. Nermut. S. Kellie. Treatise on the Cytoskeleton, Vol 4. JAI Press.

These ultrastructural studies indicated that although the microfilament bundles of these cells were disrupted if the cells were grown at the temperature permissive for pp60$^{v\text{-}src}$ tyrosine kinase activity, a large pool of disorganized filamentous actin was retained within the cells. We quantified this by assaying the polymerization status of actin in cells grown at permissive or restrictive temperature using a modified DNase I assay.[42,118] Transformed *ts*LA29Rat-1 cells contained 25% detergent-soluble actin monomer, 48% detergent-soluble polymer and 27% detergent-insoluble polymer but nontransformed *ts*LA29 cells had similar relative proportions of actin in each fraction. Therefore *ts*LA29Rat-1 cells at restrictive or permissive temperature exhibited no substantial increase in the G-actin pool or decrease in the F-actin pool despite gross changes in the actin distribution. There was also no change in the total actin content of LA29 cells at restrictive and permissive temperatures.

Those preceding experiments suggested that although the filamentous status of actin was unchanged after transformation, the ability of the filaments to bundle and interact with the plasma membrane was disrupted. Therefore we examined the interaction of microfilaments with the plasma membrane by electron microscopy. At the restrictive temperature the predominant features of such membranes were large, multifilamentous actin bundles (Fig. 3.9). These bundles contained numerous interconnections and splayed out into single filaments as they approached adhesion plaque-like structures at the membrane terminus. These membranes also contained many single filaments distributed throughout the membrane forming a network of interconnecting structures. At the permissive temperature multifilamentous bundles in these membranes were virtually absent and those which were retained contained fewer numbers of microfilaments, were shorter in length and often aggregated. In summary, our results indicated that the actin cytoskeleton in transformed cells is changed in at least three properties: structured crosslinking is reduced, actin filament lengths are shorter and membrane association is weaker. These changes must be mediated at the level of the organization of polymeric actin since there is no net depolymerization, and rather are dependent on the presence of an active pp60$^{v\text{-}src}$ protein and phosphotyrosyl proteins in the cytoskeleton. Tyrosine-specific phosphorylation of adhesion plaque proteins may account for the loss of actin-membrane association, which will be discussed in the next chapter. Although the lack of multifilamentous structures may involve a defect in actin crosslinking proteins neither of the major bundling proteins α-actinin or filamin appear to be substrates for pp60$^{v\text{-}src}$.[119] In the *ts* transformed cells described here α-actinin probably retained the capacity to interact with actin since both were present in aggregates, but could no longer effect bundling. Since capping proteins such as gelsolin shorten actin filaments without net depolymerization, it is possible that this class of protein might mediate the filament shortening we have observed, although there is no evidence of gelsolin (or any of the capping proteins) being a substrate for pp60$^{v\text{-}src}$.

The treatment of epithelial cells with a tumor promoter[120] or transformation of cells by adenovirus[121] also result in microfilament bundle disruption without any effect on actin polymerization. Furthermore, cytochalasins, which also induce rounding up, do not produce net depolymerization of actin in Hep-2 cells.[122] Whilst cytochalasin B and colchicine increase and decrease respectively the amount of G-actin in HeLa cells, other treatments such as trypsin or EDTA which induce shape changes have no effect.[123] In addition, cells can be blocked in S phase or undergo shape changes associated with mitosis without concomitant changes in

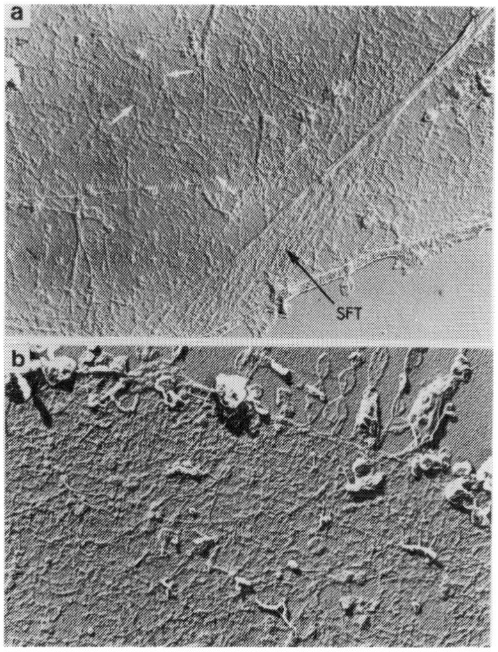

Fig. 3.9. Electron micrographs of plasma membranes of nontransformed (a) or RSV-transformed (b) fibroblasts. The cytoplasmic face of the membrane is visualized showing a stress fiber terminus interacting with the membrane in the nontransformed cell. RSV-transformed cells show no such structures. Figure courtesy of Felice et al, Eur J Cell Biol 1990; 52:47-59.

actin pool.[124] Thus, there is no general relationship between loss of microfilament bundles and the actin polymerization status, suggesting that the defect is at the level of actin crosslinking proteins, not polymerizing proteins, since actin bundling in the form of stress fibers is reduced after transformation. Changes in the biochemical characteristics of actin induced by RSV may also be partly due to transformation-specific changes in the expression of different actin isoforms or other actin-associated proteins.[125]

Not all transformed cells, however, show significant alterations in stress fiber pattern. In a study of cell lines derived from human breast tissue in vitro, Brinkley et al[126] found a wide range of actin phenotypes. In general, the absence of actin cables could be associated with rounded cell morphology and tumor cells that maintained a flattened epithelial appearance generally maintained stress fibers. In explants of human colonic epithelium the loss of actin organization generally correlated with the transition of noninvasive benign colonic tumors to invasive malignant tumors.[127]

CHANGES IN ADHESION STRUCTURES

The loss of microfilaments and cell rounding is also accompanied by disruption of other cellular structures. In particular adhesion plaques, or focal contacts, have been shown to be grossly affected by transformation. As discussed earlier, adhesion plaques are areas of close apposition of the ventral membrane of the cell with the tissue culture substratum, and are responsible for the adhesion of cells to substrata. The biochemical composition of adhesion plaques has been described elsewhere and these structures consist of accumulations of proteins which are thought to link the cytoskeleton to the plasma membrane in regions where extracellular matrix receptors (integrins) are concentrated.[2] Therefore these structures may play a role in the signaling of information from the outside of the cell to the interior, and also may participate in transduction of intracellular information to the extracellular environment.[108]

When cells are transformed by RSV there is a reduction in the number of adhesion plaques and also an alteration in the architecture of those remaining from long streaks or arrowheads to smaller, dot-like structures. Concomitantly there are also changes in the distribution of proteins associated with these structures such as α-actinin, vinculin, and talin which become localized into aggregates which appear to lie close to weak attachments on the ventral surface.[116] Compared to adhesion plaques these aggregates are highly dynamic structures capable of changing their size and location very rapidly. The rate of protein exchange in these structures is also unusually high.[109] These properties may be closely related to the changes in cellular behavior upon oncogenic transformation. Bershadsky et al[128] described two variants of focal contacts in RSV-transformed cells: dash contacts and dot contacts. Although both contained vinculin and α-actinin, only the dash contacts were associated with actin bundles. Furthermore these contacts had different intracellular sites, with dot contacts exclusively found at the active zones of the cell edge.

The aggregates of dot contacts, termed rosettes or podosomes,[129] contain many of the proteins found in focal adhesions but these proteins have a different organization. Each rosette contains a core of actin filaments together with α-actinin. The core is surrounded by vinculin and talin which can appear as rings by immunofluorescence microscopy.[93,130,131] Rosette formation does not require an extracellular matrix and perhaps that is why integrin is not usually concentrated within these abnormal adhesions.[132,133] Some normal cells such as macrophages and osteoclasts[130,131,134] also possess podosomes and the significance of these structures and their relationship to focal adhesions is not clear. However, it is interesting that these normal cells in which they are found tend to be "invasive" and express proteases at their surfaces.

In a detailed study, Gavazzi et al investigated the ultrastructure and protein constituents of podosomes in RSV-transformed BHK cells.[135] The basic structural components of podosomes were small conical bodies consisting of densely packed thin filaments. The podosomes were often joined together forming a continuous ridge, and sheets of clathrin were commonly associated. Immunoelectron microscopy revealed that vinculin occurred in podosomes either diffusely associated or in a ring surrounding the structure. Gelsolin was also associated with the podosome rosettes, as were phosphotyrosine-containing proteins and $pp60^{v\text{-}src}$. These workers speculated that the difference between podosomes and normal focal adhesions lies in the absence of some of the actin bundling proteins, or of a signal for stress fiber formation.

Other cell substratum contacts have been described in RSV-transformed cells which have different protein compositions:

point contacts which are dot-like adhesions as seen by interference reflection microscopy.[136] Although these could be found in nontransformed cells their numbers increased greatly after induction of transformation, and their composition changed from only about 20% being actin-positive to greater than 50% being actin positive. In addition most of the point contacts contained vinculin, α-actinin, talin and clathrin, and are therefore probably equivalent to, but not morphologically identical with, podosomes. Other point contacts did not contain detectable levels of these common focal contact proteins, and were positive only for clathrin and integrin. The functional role for these different substratum contacts is not known, although it is likely that since transformed cells are more motile, there has to be a balance between sufficient substrate adhesion to keep the cells anchored but weak enough adhesions to allow the cells to move. The small, less adherent and more transient point contacts may fulfill this role. The small aggregates of F-actin seen by immunofluorescence and the appearance of these interconnections correlates with a decrease in the number of stable adhesion structures. These aggregates do not appear to associate strongly with the plasma membrane since they are lost during the preparation of substrate-adherent ventral membranes, consistent with a report that transformed cells have 30% less actin associated with the membrane than nontransformed cells.[137] The single filaments which are retained in the substrate-adherent membranes are commonly much shorter than in nontransformed cells.

Changes in other adhesion structures have also been demonstrated in RSV-transformed cells. The use of a temperature-sensitive RSV mutant[138,139] showed that induction of transformation in chicken lens cells leads to a reduction in intercellular adherens-type junctions and a major reorganization of the transmembrane junctional protein N-cadherin. Although N-cadherin is not phosphorylated on tyrosine in these cells, some associated proteins are, and this will be discussed in a later section. Behrens et al[139] introduced a similar *ts src* into MDCK cells and observed a rapid loss of cell-cell contacts. In these cells E-cadherin was phosphorylated on tyrosine. It is likely that tyrosine phosphorylation regulates N-cadherin organization, since stimulation of tyrosine phosphorylation in nontransformed cells by vanadyl hydroperoxide or inhibition of tyrosine phosphorylation by tyrphostins in transformed cells results in deterioration or restoration of adherens junctions respectively.[139,140]

CHANGES IN THE EXPRESSION OF CYTOSKELETAL PROTEINS IN TRANSFORMED CELLS

The mechanism by which there is a decrease in the number of stress fibers in transformed cells is still unknown although there are a number of possibilities. As already discussed, a major change is in the amount of actin bundling or crosslinking rather than polymerization. Whilst this might be a consequence of direct phosphorylation of a cytoskeletal component catalyzed by $pp60^{v\text{-}src}$ it is possible that the cytoskeletal changes are due to an indirect effect of $pp60^{v\text{-}src}$ on the expression of cytoskeletal proteins by, for example, differential activation of specific transcription factors. Several changes in actin and its associated proteins have been detected in a variety of transformed cells and these are discussed below.

ACTIN

Differential expression of actin isoforms in transformed cells has been reported, for example α-actin is reduced in several transformed cell types.[141] An increase in G actin has been reported in transformed rat fibroblasts and differences in β-actin have also been observed.[142] Since different isoforms have different polymerization properties and binding affinities for associated proteins changes in these proteins may lead to differential actin organization in transformed cells.[143,144]

Tropomyosin

Three major tropomyosin isoforms are normally expressed in fibroblasts, however after transformation by retroviruses such as RSV the total amount of tropomyosin is reduced and there is differential expression such that a higher molecular weight isoform is downregulated with respect to the other isoforms.[145,146] The functional relevance of this has been emphasized by the experiments showing that transfection of transformed cells with tropomyosin isoforms can partially revert the transformed phenotype.[147] It is possible that the changes in tropomyosin isoform expression might lead to lack of stability of actin filaments, making the filaments more susceptible to other actin-associated proteins.

Gelsolin

Gelsolin is an abundant protein which regulates actin assembly in calcium-dependent manner. The amount of gelsolin in transformed cells is reduced compared to normal counterparts and there is a differential distribution of phosphorylated gelsolin in transformed cell.[148] Alterations in calcium in transformed cells may also affect the function of gelsolin.

Caldesmon

Caldesmon stabilizes stress fibers and its binding to polymeric actin is increased by the interaction with tropomyosin in a calcium-calmodulin dependent manner. There is a differential distribution of caldesmon from ruffling membranes to a diffuse localization following transformation by RSV.[149] The caldesmon content can be reduced in other transformed cells however it is not clear whether the change in caldesmon localization is due to a lack of actin bundles or whether changes in caldesmon result in changes in actin configuration.

MARCKS

MARCKS is a major substrate for protein kinase C and its potential role in actin changes will also be discussed in a later section. Cells transformed by *v-src* express a reduced level of MARCKS, and this protein is also phosphorylated. A potential interrelationship between MARCKS and tyrosine phosphorylation in actin regulation was also revealed by experiments in which the inhibition of tyrosine kinase activity of pp60$^{v\text{-}src}$ led to a restoration of MARCKS protein expression to levels found nontransformed cells.[150,151]

Cytoskeletal Changes Associated with Activation of Small GTP-Binding Proteins

Many tumors contain mutations in the *ras* gene, and the physiological role of p21^{ras} and its related proteins have been the subject of intense investigation (for review see Bokoch and Der[152]). A role for *ras* in the disruption of the cytoskeleton was suggested on the basis that *ras*-transformed cells were morphologically different from nontransformed cells, and that microinjection of oncogenic *ras* proteins into cells resulted in proliferation and morphological changes.[153] Recently, two *ras*-related proteins, *rho* and *rac*, have been implicated in regulating the actin skeleton in normal cells. Paterson et al[154] showed that microinjection of *rho* proteins into cells resulted in a rapid reorganization of actin stress fibers. Ridley and Hall[155] have extended these studies to show that *rho* also induced vinculin and talin-containing focal adhesions. The induction of stress fiber formation by lysophosphatidic acid or growth factors in serum-starved fibroblasts was inhibited by microinjecting Chlostridium botulinum C3 transferase, which inactivates *rho* by ADP ribosylation. Whilst stress fiber formation was dependent on an active *rho* protein, membrane ruffling was unaffected. However microinjection of *rac1* into serum-starved fibroblasts stimulated actin filament accumulation at the plasma membrane, membrane ruffling and pinocytosis. Inactivation of *rac1* inhibited growth factor and oncogenic *ras*-induced membrane ruffling.[156] Thus regulation of actin by growth factors is mediated via *rho* and *rac* proteins. By analogy with the effects of growth factors described, it is possible that these GTP-binding proteins are directly affected oncogenic tyrosine kinases. In support of this recently two proteins have been identified which couple tyrosine phosphorylation to *ras* activation. Sos is an SH2 domain-containing protein which possesses guanine nucleotide releasing activity for H-*ras*, and this is enhanced by associating with another protein, Grb2, which contains both SH2 and SH3 domains.[157,158,159] Transformation of cells by v-*src* results in tyrosine phosphorylation of Shc, another SH2 domain-containing protein. This in turn leads to formation of a complex between Shc, Grb2 and Sos which may stimulate Sos-mediated guanine nucleotide exchange on *ras* (reviewed in refs 160 and 161). It may be that there are similar proteins involved in the enhancement of nucleotide exchange on *rac1* and *rho*, thus coupling *src*-induced tyrosine phosphorylation to rearrangement of intracellular actin by the activation of small GTP-binding proteins.

References

1. Kellie S, Horvath AR, Elmore MA. Cytoskeletal targets for oncogenic tyrosine kinases. J Cell Sci 1990; 99: 207-211.
2. Burridge, K. Substrate adhesions in normal and transformed fibroblasts: organization and regulation of cytoskeletal, membrane and extracellular matrix components. Cancer Rev 1986; 4: 18-78.
3. Heasman JEM. Contact inhibition of locomotion: a reappraisal. Int Rev Cyt 1978; 55: 49-66
4. Stoker M, O'Neill C, Berryman S et al. Anchorage and growth regulation in normal and virus-transformed cells. Int J Cancer 1968; 3: 683-693.
5. Temin HM, Rubin H. Characteristics of an assay for Rous sarcoma virus and Rous sarcoma cells in tissue culture. Virol 1958; 6: 669-688
6. Bissell MJ. The differentiated state of normal and malignant cells or how to define a "normal" cell in culture. Int Rev Cytol 1984; 70: 27-100.
7. O'Neill C, Jordan P, Riddle P. Narrow linear strips of adhesive substratum are powerful inducers of both growth and total focal contact area. J Cell Sci 1990; 95: 577-586.
8. Raz A, Ben Ze'ev A. Growth control and cell spreading: differential response in preneoplastic and metastatic cell variants. Int J Cancer 1982; 29: 711-715
9. Ben-Ze'ev A. Animal cell shape changes and gene expression. BioEssays 1991; 13: 207-212.
10. Ben-Ze'ev A. Regulation of cytoskeletal protein synethesis in normal and cancer cells. Cancer Rev 1986; 4: 91-115.
11. Hynes RO. Alteration of cell-surface proteins by viral transformation and by proteolysis. Proc Natl Acad Sci USA 1973; 70: 3170-3174
12. Hynes RO. Cell surface protein and malignant transformation. Biochim Biophys Acta 1976; 458: 73-107.
13. Olden K, Yamada KM. Mechanism of the decrease in the major cell surface protein of chick embryo fibroblasts after transformation. Cell 1977; 11: 957-969.
14. Yamada KM, Yamada SS, Pastan I. Cell surface protein partially restores morphology, adhesiveness and contact inhibition of movement to transformed fibroblasts. Proc Natl Acad Sci USA 1976; 73: 1217-1221.
15. Ali IU, Mautner V, Lanza R, Hynes RO. Restoration of normal morphology, adhesion and cytoskeleton in transformed cells by addition of a transformation-sensitive surface protein. Cell 1977; 11: 115-126.
16. Kellie S, Patel B, Wigglesworth NM et al. The use of Rous sarcoma virus transformation mutants with differing tyrosine kinase activities to study the relationships between vinculin phosphorylation, pp60$^{v\text{-}src}$ location and adhesion plaque integrity. Exp Cell Res 1986; 165: 216-228.
17. Kellie S, Patel B, Wigglesworth, NM et al. Comparison of the relative importance of tyrosine-specific vinculin phosphorylation and the loss of surface-associated fibronectin in the morphology of cells transformed by Rous sarcoma virus. J Cell Sci 1986; 82: 129-142.
18. Willingham MC, Yamada KM, Yamada, SS et al. Microfilament bundles and cell shape are related to adhesiveness to substratum and are dissociable from growth control in

cultures fibroblasts. Cell 1977; 10: 375-380.
19. Chen WT, Hasegawa E, Hasegawa T et al. Development of cell surface linkage complexes in cultured fibroblasts. J Cell Biol 1985; 100: 1103-1114.
20. Unkeless JC, Tobia A, Ossowski L et al. An enzymatic function associated with transformation of fibroblasts by oncogenic viruses. I Chick embryo fibroblast cultures transformed by avian RNA tumor viruses. J Exp Med 1973; 137: 85-111.
21. Chen JM, Chen WT. Fibronectin-degrading proteases from the membranes of transformed cells. Cell 1987; 48: 198-203.
22. Chen WT, Olden K, Bernard BA, Chu F. Expression of transformation-associated proteases that degrade fibronectin at cell contact sites. J Cell Biol 1984; 98: 1546-1555.
23. Chen WT, Chen JM, Parsons SJ, Parsons, JT. Local degradation of fibronectin at sites of expression of the transforming gene product pp60$^{v\text{-}src}$. Nature (London) 1985; 316: 156-158.
24. Burger MM. Proteolytic enzymes initiating cell division and escape from contact inhibition of growth. Nature (London) 1970; 227: 170-171
25. Sefton BM, Rubin H. Release from density-dependent inhibition of growth by proteolytic enzymes. Nature (London) 1970; 227: 843-845.
26. Pollack R, Rifkin D. Actin-containing cables within anchorage-dependent rat embryo cells are dissociated with plasmin and trypsin. Cell 1975; 6: 495-506.
27. Goldberg AR. Involvement of cell and serum proteases in exposing tumor-specific agglutination sites. Ann NY Acad Sci 1974; 234: 348-354.
28. Hatcher VB, Oberman MS, Wertheim MS et al. The relationship between surface protease activity and the rate of cell proliferation in normal and transformed cells. Biochim Biophys Acta 1977; 76: 602-608.
29. Chou IN, Black PH, Roblin RO. Non-selective inhibition of transformed cell growth by a protease inhibitor. Proc Natl Acad Sci USA 1974; 71: 1748-1752.
30. Weber MJ. Inhibition of protease activity in cultures of Rous sarcoma virus-transformed cells: effect on the transformed phenotype. Cell 1975; 5: 253-261.
31. Goldberg AR. Increased protease levels in transformed cells: a casein overlay assay for the detection of plasminogen activator production. Cell 1974; 2: 95-102.
32. Quigley JP. Phorbol ester-induced morphological changes in transformed chick fibroblasts: evidence for direct catalytic involvement of plasminogen activator. Cell 1979: 17: 131-141.
33. Fairburn S, Gilbert R, Ojakian G et al. The extracellular matrix of normal chick embryo fibroblasts: its effect on transformed chick fibroblasts and its proteolytic degradation by the transformants. J Cell Biol 1985; 101: 1790-1798.
34. Pollanen J, Saksela O, Salonen EM et al. Distinct localizations of urokinase-type plasminogen activator and its inhibitor under cultured human fibroblasts and sarcoma cells. J Cell Biol 1977; 104: 1085-1096.
35. Pollanen J, Hedman K, Neilsen LS et al. Ultrastructural localization of plasma membrane-associated urokinase-type plasminogen activator at focal contacts. J Cell Biol 1988; 106: 87-95.
36. Hebert CA, Baker JB. Linkage of extracellular plasminogen activator to fibroblast cytoskeleton: colocalization of cell surface urokinase with vinculin. J Cell Biol 1988; 106: 1241-1247.
37. Ossowski L, Quigley JP, Reich E. Fibrinolysis associated with oncogenic transformation. Morphological correlates. J Biol Chem 1974; 249: 4312-4316.
38. Pollak R, Risser R, Conlon S, Rifkin D. Plasminogen activator production accompanies loss of anchorage regulation in transformation of primary rat embryo cells by Simian virus 40. Proc Natl Acad Sci USA 1974; 71: 4792-4796.
39. McNutt NS, Culp LA, Black PH. Contact-inhibited revertant cell lines isolated from SV40-transformed cells II. Ultrastructural study. J Cell Biol 1971; 50: 691-708.
40. McNutt NS, Culp LA, Black PH. Contact-inhibited revertant cell lines isolated from SV40-transformed cells IV. Microfilament distribution and cell shape change in nontransformed, transformed and revertant Balb/c 3T3 cells. J Cell Biol 1973; 56:

412-428.
41. Wang EA, Goldberg R. Changes in microfilament organization and surface topography upon transformation of chick embryo fibroblasts with Rous sarcoma virus. Proc Natl Acad Sci USA 1976; 73: 4065-4069.
42. Felice GR, Eason P, Nermut MV, Kellie S. pp60^{v-src} association with the cytoskeleton induces actin reorganization without affecting polymerization status. Eur J Cell Biol 1990; 52: 47-59.
43. Wang Y-L. Exchange of actin subunits at the leading edge of living fibroblasts: possible role in treadmilling. J Cell Biol 1985; 101: 597-602.
44. Pollard TD, Cooper JA. Actin and actin binding proteins. Ann Rev Biochem 1986; 55: 987-1035.
45. Sanger JW, Sanger JM. Banding and polarity of actin filaments in interphase and cleaving cells. J. Cell Biol 1980; 86: 568-575.
46. Byers HR, Fujiwara K. Stress fibers in situ. Immunofluorescent visualization with antiactin, antimyosin and anti-alpha-actinin. J Cell Biol 1982; 93: 804-811.
47. Goldman RD, Talian J, Goldman A et al. Adhesion plaques of embryonic fibroblasts are analagous to Z-lines. J Cell Biol 1979 83: 1748a.
48. Weber K, Groeschel-Stewart U. Antibody to myosin: The specific visualization of myosin-containing filaments in nonmuscle cells. Proc Natl Acad Sci USA 1974; 71: 4561-4564.-771.
49. Lazarides E. Actin, alpha-actinin and tropomyosin interaction in the structural organization of actin filamants in non-muscle cells. J Cell Biol 1975; 65: 549-561.
50. Wang E, Yin HL, Krueger JG et al. Unphosphorylated gelsolin is localized in regions of cell-substratum contact or attachment in Rous sarcoma virus-transformed cells. J Cell Biol 1984; 98: 761-771.
51. Lazarides E, Burridge K. Alpha-actinin: immunofluorescent colocalization of a muscle structural protein in non-muscle cells. Cell 1975; 6: 289-298.
52. Couchman JR, Rees DA.The behaviour of fibroblasts migrating from chick heart explants. Changes in adhesion, locomotion, growth and in the distribution of actomyosin and fibronectin. J Cell Sci 1979; 39: 149-65.
53. Couchman JR, Rees DA, Green MR et al. Fibronectin has a dual role in locomotion and anchorage of primary chick fibroblasts and can promote entry into the division cycle. J. Cell Biol 1982; 93: 402-410.
54. Mooseker MS, Bonder, EM, Conzelman KA et al. Brush border cytoskeleton and integration of cellular function. J Cell Biol 1984; 99: 1045-1125.
55. Mooseker MS. Organization, chemistry and assembly of the cytoskeletal apparatus of the intestinal brush border. Ann Rev Cell Biol 1985; 1: 261-293.
56. Bennett V, Gilligan DM. The spectrin-based membrane skeleton and micron-scale organization of the plasma membrane. Ann Rev Cell Biol 1993; 9: 27-66.
57. Jones P, Jackson P, Price GJ et al. Identification of a talin binding site in the cytoskeletal protein vinculin. J Cell Biol 1989; 109:2917-2927.
58. Otey C, Pavalko FM, Burridge K. An interaction between α-actinin and the β1 integrin subunit in vitro. J Cell Biol 1990; 11:721-729.
59. Nicol A, Nermut MV. A new type of substratum adhesion structure in NRK cells revealed by correlated interference reflection and electron microscopy. Eur J Cell Biol 1987; 43: 348-357.
60. Painter RG, Prodouz KN, Gaarde E. Isolation of a subpopulation of glycoprotein IIb-IIIa from platelet membranes that is bound to actin. J Cell Biol 1985; 100: 652-657.
61. Molony L, Burridge K. Molecular shape and self association of vinculin and metavinculin. J Cell Biochem 1985; 29: 31-36.
62. Bush WA, Stromer MH, Goll DE et al. Calcium-specific removal of Z-lines from rabbit skeletal muscle. J Cell Biol 1972; 52: 367-381.
63. Burridge K, McCullough L. The association of of alpha-actinin with the plasma membrane. J Supramol Structure 1980; 13: 53-65.
64. Davison MD, and Critchley DR. Alpha-actinin and the DMD protein contain spectrin-like repeats. Cell 1988; 52: 159-160.

65. Blanchard A, Ohanian V, Critchley DR. The structure and function of alpha-actinin. J Muscle Res Cell Motil 1989; 10: 280-289.
66. Geiger B. A130k protein from chicken gizzard: its localization at the termini of microfilament bundles in cultured chicken cells. Cell 1979; 18: 193-205.
67. Geiger B, Tokuyasu KT, Dutton AH et al. Vinculin, an intracellular protein localized at specialized sites where microfilament bundles terminate at cell membranes. Proc Natl Acad Sci USA 1980; 77: 4127-4131.
68. Geiger B. Microheterogeneity of avian and mammalian vinculin. Distinctive intracellular distribution of different isovinculins. J Mol Biol 1982; 159: 685-701.
69. Feramisco JR, Smart JE, Burridge K et al. Co-existence of vinculin and a vinculin-like protein of higher molecular weight in smooth muscle. J Biol Chem 1982; 257: 11024-11031.
70. Siliciano J, Craig S. Metavinculin—a vinculin-related protein with the solubility properties of a membrane protein. Nature (London)1982; 330: 533-535.
71. Jockusch B, Isenberg G. Interaction of alpha-actinin and vinculin with actin: opposite effects on filament network formation. Proc Natl Acad Sci USA 1981; 78: 3005-3009.
72. Wilkins JA, Lin S. High affinity interaction of vinculin with actin filaments in vitro. Cell 1982; 28: 83-90.
73. Evans RR, Robson RM, Stromer MH. Properties of smooth muscle vinculin. J Biol Chem 1984; 259: 3916-3924.
74. Schroer E, Wegner A. Purification and characterization of a protein from chicken gizzard which inhibits actin polymerization. Eur J Biochem 1985; 153: 515-520.
75. Otto JJ. Detection of vinculin-binding proteins with an ^{125}I-vinculin gel overlay. J Cell Biol 1983; 97: 1283-1287.
76. Ruhnau K, Wegner A. Evidence for direct binding of vinculin to actin filaments. FEBS Letts 1988; 228: 105-108.
77. Avnur Z, Small V, Geiger B. Actin-independent association of vinculin with the cytoplasmic aspect of the plasma membrane in cell-cell contact areas. J Cell Biol 1983; 96: 1622-1630.
78. Burridge K. Substrate adhesions in normal and transformed fibroblasts: organization and regulation of cytoskeleton, membrane and extracellular matrix components of focal contacts. Cancer Rev 1986; 4: 18-78.
79. Volberg T, Geiger B, Kartenbeck J et al. Changes in membrane-microfilament interaction in intercellular adherens junctions upon removal of calcium ions. J Cell Biol 1986; 102: 1832-1842.
80. Westmeyer A, Ruhnau, Wegner A et al. Antibody mapping of functional domains in vinculin. EMBO J. 1990; 9: 2071-2078.
81. Price GJ, Jones P, Davison MD et al. Isolation and characterization of a vinculin cDNA from chick embryo fibroblasts. Biochem J 1987; 245: 595-603.
82. Singer II, Paradiso PR. A transmembrane relationship between fibronectin and vinculin (130kD protein): serum modulation in normal and transformed fibroblasts. Cell 1981; 24: 481-492.
83. Hynes RO, Destree AT. Relationship between fibronectin (LETS protein) and actin. Cell 1978; 15: 875-886.
84. Heggeness MH, Wang K, Singer SJ. Intracellular distributions of mechanochemical proteins in cultured fibroblasts. Proc Natl Acad Sci USA 1977; 74: 3883-3887.
85. Singer II. The fibronexus: a transmembrane association between fibronectin-containing fibers and bundles of 5nm filaments in human fibroblasts. Cell 1979 16: 675-685.
86. DePasquale J, Izzard CS. Evidence for an actin-containing precursor of the focal contact and the timing of incorporation of vinculin at the focal contact. J Cell Biol 1987; 105: 2803-2809.
87. Burn P, Burger MM. The cytoskeletal protein contains transformation-sensitive, covalently bound lipid. Science 1987; 235: 476-479.
88. Kellie S, Wigglesworth NM. The cytoskeletal protein vinculin is acylated by myristic acid. FEBS Letts 1987; 213: 428-432.
89. Niggli V, Dimitrov VP, Brunner J et al. Interaction of the cytoskeletal protein vinculin with bilayer structures analyzed with a photoactivatable phospholipid. J Biol Chem. 1986; 261: 6912-6918.
90. Burridge K, Mangeat P. An interaction

between vinculin and talin. Nature (London) 1984; 308: 744-746.

91. Jones P, Jackson P, Price GJ et al. Identification of a talin binding site in the cytoskeletal protein vinculin. J Cell Biol 1989; 109: 2917-2927.
92. Price GJ, Jones P, Davison MD et al. Primary sequence and domain structure of chicken vinculin. Biochem J 1989; 259: 453-461.
93. Burridge K, Connell L. A new protein of adhesion plaques. J Cell Biol 1983; 97: 359-367.
94. Horwitz A, Duggan K, Buck C et al. Interaction of plasma membrane fibronectin receptor with talin - a transmembrane linkage. Nature (London) 1986; 320; 531-533.
95. Muguruma M, Matsumura S, Fukazawa T. Direct interactions between talin and actin. Biochem Biophys Res Comm. 1990; 171: 1217-1223.
96. Nuckolls GH, Turner CE, Burridge K. Functional studies of the domains of talin. J Cell Biol 1990; 110: 1635-1644.
97. Brands R, De Boer A, Feltkamp CA, Roos E. Disintegration of adhesion plaques in chick embryo fibroblasts upon Rous sarcoma virus transformation: different dissociation rates for talin and vinculin. Exp Cell Res 1990; 186: 138-148.
98. Hynes RO. Integrins: a family of cell surface receptors. Cell 1987; 48: 549-554.
99. Nermut MV, Green NM, Eason P et al. Electron microscopy and structural model of human fibronectin receptor. EMBO J 1988; 7: 4093-4099.
100. Pytela R, Pierschbacher MD, Ruoslahti E. Identification and isolation of a 140kd cell surface glycoprotein with properties expected of a fibronectin receptor. Cell 1985; 40:191-198.
101. Ruoslahti E, Pierschbacher MD. New perspectives in cell adhesion: RGD and integrins. Science 1987; 491-497.
102. Singer II, Scott S, Kawka DW et al. Cell-surface distribution of fibronectin and vitronectin receptors depends on substrate composition and extracellular matrix accumulation. J Cell Biol 1988; 106: 2171-2182.
103. Saga S, Chen W, Yamada KM. Enhanced fibronectin receptor expression in Rous sarcoma virus-induced tumors. Cancer Res 1988; 48: 5510-5513.
104. Akiyama SK, Larjava H, Yamada KM. Differences in the biosynthesis and localization of the fibronectin receptor in normal and transformed cultured human cells. Cancer Res 1990; 50: 1601-1607.
105. McDonald JA, Quade BJ, Broekelmann TJ et al. Fibronectin's cell adhesion domain and an amino-terminal matrix assembly domain participate in its assembly into fibroblast pericellular matrix. J Biol Chem 1987; 272: 2957-2967.
106. Buck CA, Horwitz AF. Cell surface receptors for extracellular matrix molecules. Ann Rev Cell Biol 1987; 3: 179-206.
107. Roman J, LaChance RM, Broekelmann TJ et al. The fibronectin receptor is organized by the exracellular matrix fibronectin: implications for oncogenic transformation and for cell recognition of fibronectin matrices. J Cell Biol 1989; 108: 2529-2543.
108. Humphries MJ, Mould AP, Tuckwell, DS. Dynamic aspects of adhesion receptor function - integrins both twist and shout. BioEssays 1993; 15: 391-397.
109. Stickel SK, Wang Y. Alpha-actinin-containing aggregates in transformed cell are highly dynamic structures. J Cell Biol 1987; 104: 1521-1526.
110. Argraves WS, Dickerson K, Burgess WH et al. Fibulin, a novel protein that interacts with the fibronectin receptor B subunit cytoplasmic domain. Cell 1989; 58: 623-629.
111. Pollack R, Osborn M, Weber K. Patterns of organization of actin and myosin in normal and transformed cells. Proc Natl Acad Sci. USA 1975; 72: 994-998.
112. Wyke JA, Stoker A. Genetic analysis of the form and function of the viral *src* gene. Biochim Biophys Acta 1987; 907: 47-70.
113. Ambros VR, Chen LB, Buchanan JM. Surface ruffles as markers for studies of cell transformation by Rous sarcoma virus. Proc Natl Acad Sci USA 1975; 72: 3144-3148.
114. Edelman GM, Yahara I. Temperature-sensitive changes in surface modulation assemblies of fibroblasts transformed by mutants of Rous sarcoma virus. Proc Natl Acad Sci USA 1976; 73: 2047-2051.

115. Boschek CB, Jockusch BM, Friis RR et al. Early changes in the distribution and organization of micro-filament proteins during cell transformation. Cell 1981; 24: 175-184.
116. David-Pfeuty T, Singer SJ. Altered distributions of the cytoskeletal proteins vinculin and alpha-actinin in cultured fibroblasts transformed by Rous sarcoma virus. Proc Natl Acad Sci USA 1980; 77: 6687-6691.
117. Rubin RW, Warren RH, Lukemann RS, Clements, A. Actin content and organization in normal and transformed cells in culture. J Cell Biol 1978; 78, 28-35.
118. Holme TC, Kellie S, Wyke JA et al. Effect of transformation by Rous sarcoma virus on the character and distribution of actin in Rat-1 fibroblasts: A biochemical and microscopical study. Brit J Cancer 1986; 53: 465-476.
119. Sefton BM, Hunter T, Ball EH, Singer SJ. Vinculin: A cytoskeletal target of the transforming protein of Rous sarcoma virus. Cell 1981; 24: 165-174.
120. Kellie S, Holme TC, Bissell MJ. Interaction of tumour promoters with epithelial cells in culture. Exptl Cell Res 1985; 160: 259-274.
121. Jackson P, Bellett, AJD. Reduced microfilament organization in adenovirus type 5-infected rat embryo cells: a function of early region 1a. J Virol 1985; 55: 644-650.
122. Morris A, Tannenbaum J. Cytochalasin D does not produce net depolymerization of actin filaments in HEp-2 cells. Nature (London) 1980; 287: 637-639.
123. Blikstad I, Carlsson L. On the dynamics of the microfilament system in HeLa cells. J Cell Biol 1982; 93: 122-128.
124. Heacock CS, Bambourg SR. Levels of filamentous and globular actin in Chinese Hamster ovary cells throughout the cell cycle. Exptl Cell Res 1983; 147: 240-246.
125. Witt DP, Brown DP, Gordon J. Transformation-sensitive isoactin in passaged chick embryo fibroblasts transformed by Rous sarcoma virus. J Cell Biol 1983; 96: 1766-1771.
126. Brinkley BR, Beall P, Wible L et al. Variations in cell form and cytoskeleton in human breast carcinoma cells in vitro. Cancer Res 1980; 40: 3118-3129.
127. Friedman E, Verderame M, Winawer S, Pollack R. Actin cytoskeletal reorganization loss in benign-to-malignant tumor transitions in cultured human colonic cells. Cancer Res 1984; 44: 3040-3050.
128. Bershadsky AD, Tint IS, Neyfahk A et al. Focal contacts of normal and RSV-transformed quail cells. Exptl Cell Res 1985; 158: 433-444.
129. Tarone G, Cirillo D, Giancotti FG et al. Rous sarcoma virus-transformed fibroblasts adhere primarily at discrete protrusions of the ventral membrane called podsomes. Exptl Cell Res 1985; 159: 141-157.
130. Marchisio PC, Cirilo D, Naldini L et al. Cell-substratum interaction of cultured avian osteoclasts is mediated by specific adhesion structures. J Cell Biol 1984; 99: 1696-1705.
131. Marchisio PC., Cirillo D, Teti A et al. Rous sarcoma virus-transformed fibroblasts and cells of monocytic origin display a peculiar dot-like organization of cytoskeletal proteins involved in microfilament-membrane interactions. Exptl Cell Res 1987; 169: 202-214.
132. Chen W-T, Wang J, Hasegawa T et al. Regulation of fibronectin receptor distribution by transformation, exogenous fibronectin and synthetic peptides. J Cell Biol 1986; 103: 1649-1661.
133. Giancotti FG, Ruoslahti E. Elevated levels of $\alpha_5\beta_1$ fibronectin receptor suppresses the transformed phenotype of chinese hamster ovary cells. Cell 1990; 69: 849-859.
134. Lehto VP, Hori T, Varti T et al. Reorganization of cytoskeletal and contractile elements during transition of human monocytes into adherent macrophages. Lab Invest 1982; 47: 391-399.
135. Gavazzi I, Nermut MV, Marchisio PC. Ultrastructure and gold immunolabelling of cell-substratum adhesions (podosomes) in RSV-transformed BHK cells. J Cell Sci 1989; 94: 85-99.
136. Nermut MV, Eason P, Hirst EM, Kellie S. Cell/substratum contacts in RSV-transformed cells. Exptl Cell Res 1988; 193: 382-387.
137. Wickus G, Gruenstein E, Robbins PW et al. Decrease in membrane-associated actin of fibroblasts after transformation by Rous sarcoma virus. Proc Natl Acad Sci USA 1975; 72: 746-749.
138. Volberg T, Geiger B, Dror R, Zick Y.

Modulation of intercellular adherens-type junctions and tyrosine phosphorylation of their components in RSV-transformed chick lens cells. Cell Reg 1991; 2: 105-120.

139. Behrens J, Vakaet L, Friis R et al. Loss of epithelial differentiation and gain of invasiveness correlates with tyrosine phosphorylation of the E-cadherin/β-catenin complex in cell transformed with a temperature-sensitive v-src gene. J Cell Biol 1993; 120: 757-766.

140. Volberg T, Zick Y, Dror R et al. The effect of tyrosine-specific phosphorylation on the assembly of adherens-type junctions. EMBO J 1992; 11: 1733-1742.

141. Leavitt J, Gunning P, Kedes L et al. Smooth muscle alpha actin is a transformation-sensitive marker for mouse NIH3T3 and Rat-2 cells. Nature (London) 1985; 316: 840-842.

142. Leavitt J, Kakunagas T. Expression of a variant form of actin and additional changes following chemical-induced in vitro neoplastic transformation of human fibroblasts. J Biol Chem 1980; 255: 1650-1661.

143. Carlier M-F. Actin: protein structure and filament dynamics. J Biol Chem 1991; 266: 1-4.

144. Vanderkerckhove J. Actin binding proteins. Curr Opinion Cell Biol 1990; 2: 41-50.

145. Prasad GL, Fulder RA, Cooper HL. Expression of transduced tropomyosin 1 cDNA suppresses neoplastic growth of cells transformed by ras oncogene. J Cell Biol 1990; 90: 7039-7043.

146. Matsumara F, Lin JJC, Yamashiro-Matsumura S et al. Differential expression of tropomyosin forms in the microfilaments isolated from normal and transformed rat cultured cells. J Biol Chem 1983; 258: 13954-13964.

147. Leavitt J, Latter G, Lutomski L et al. Tropomyosin isoform switching in tumorigenic human fibroblasts. Mol Cell Biol 1986; 6: 2721-2726.

148. Vanderkerkhove J, Bauw G, Vancompernelle K et al. Comparative two dimensional gel analysis and microsequencing identifies gelsolin as one of the most prominent downregulated markers of transformed human fibroblasts and epithelial cells. J Cell Biol 1990; 111: 95-102.

149. Owada MK, Hakura A, Ida K et al. Occurence of caldesmon (a calcium-binding protein) in cultured cells: comparison of normal and transformed cells. J Cell Biol 1984; 81: 3133-3137.

150. Joseph CK, Qureshi SA, Wallace DJ et al. MARCKS protein is transcriptionally downregulated in v-src transformed BALB/c 3T3 cells. J Biol Chem 1992; 267: 1327-1330.

151. Wolfman A, Wingrove TG, Blackshear PJ et al. Downregulation of protein kinase C and of an endogenous 80-kDa substrate in transformed fibroblasts. J Biol Chem 1987; 262: 16546-16552.

152. Bokoch GM, Der CJ. Emerging concepts in the ras superfamily of GTP-binding proteins. FASEB J 1993; 7: 750-759.

153. Feramisco JR, Kamata T, Gross M et al. Microinjection of the oncogenic form of the human H-ras (T24) protein results in rapid proliferation of quiescent cells. Cell 1984; 38: 109-117.

154. Paterson HF, Self AJ, Garrett MD et al. Microinjection of recombinant $p21^{rho}$ induces rapid changes in cell morphology. J Cell Biol 1990; 111: 1001-1007.

155. Ridley AJ, Hall A. The small GTP-binding protein rho regulates the assembly of focal adhesions and actin stress fibers in response to growth factors. Cell 1992; 70: 389-399.

156. Ridley AJ, Paterson HF, Johnston, CL et al. The small GTP-binding protein rac regulates growth factor-induced membrane ruffling. Cell 1992; 70: 401-410.

157. Gale WN, Kaplan D, Lowenstein E et al. GRB2 mediates the EGF-dependent activation of guanine nucleotide exchange on ras. Nature (London) 1993; 363: 88-92.

158. Rosakis-Adcock M, McGlade J, Mbamalu G et al. Association of shc and GRb2/sem-5 SH2 containing proteins is implicated in the activation of ras pathway by tyrosine kinases. Nature (London) 1992; 360: 689-692.

159. Buday L, Downward J. Epidermal growth factor regulates p21*ras* through the formation of a complex of receptor, Grb2 adaptor protein and Sos nucleotide exchange factor. Cell 1993; 73: 611-620.

160. Panayotou G, Waterfield MW. The assembly of signaling complexes by receptor tyrosine kinases. BioEssays 1993; 15: 171-177.

161. Pawson T, Schlessinger J. SH2 and SH3 domains. Current Biology 1993; 3: 434-442.

CHAPTER 4

MOLECULAR MECHANISMS UNDERLYING *V-SRC* TRANSFORMATION

THE DISTRIBUTION OF pp60$^{V\text{-}SRC}$ IN TRANSFORMED CELLS

It is now well established that various oncogenes localize in different places within the cell. For example, SV40 large T antigen, *jun* and *fos* are concentrated in the nucleus,[1,2,3] whereas others such as *fms* and *fps* are found in the cytoplasm or membranes.[4,5] Immunofluorescent and electron microscopical studies initially revealed that pp60$^{v\text{-}src}$ was localized in the cytoplasm, with some additional indication of concentrations at the plasma membrane.[6] This was consistent with biochemical analyses, since when RSV-transformed cells were homogenized or broken apart the majority of pp60$^{v\text{-}src}$ remained associated with the particulate fraction of the cells, although a minority was also found to be cytosolic.[7] As discussed in chapter 2 it is now recognized that myristoylation of the N-terminal glycine residue of pp60$^{v\text{-}src}$ is essential for this membrane association, although there may be other requirements since in some *ts* mutants such as *ts*LA29 pp60$^{v\text{-}src}$ is myristoylated at the restrictive temperature but remains cytosolic.[8] The investigation of other mutant proteins has demonstrated a prerequisite for plasma membrane localization before efficient transformation can occur.[9,10,11] Recently Kaplan et al[12] have described three domains which are capable of mediating the attachment of pp60$^{v\text{-}src}$ to different subcellular membranes, and a 32 kDa plasma membrane protein which binds to myristoylated pp60$^{v\text{-}src}$ has also been identified.[13]

THE LOCALIZATION OF PP60$^{V\text{-}SRC}$ TO ADHESION PLAQUES

In a detailed immunofluorescence study of RSV-transformed chick embryo fibroblasts pp60$^{v\text{-}src}$ was found to be concentrated in the remnant adhesion plaques and related structures such as rosettes and podosomes where it co-localized with other adhesion plaque proteins such as vinculin, talin and paxillin.[14] This distribution coincided with staining for phosphotyrosine-containing proteins, which were also concentrated in the remnant adhesion structures.[15] Although the association of pp60$^{v\text{-}src}$ with podosomes was apparent in many transformed cell types, is this necessary for transformation? To address this question morphological mutants of RSV have been utilized which allow careful examination of transformation parameters, in particular the effect of the presence of $pp60^{v\text{-}src}$ on adhesion plaque integrity, cell morphology and cytoskeletal organization. The work of a number of groups is summarized in Table 4.1. Many mutants showed a correlation between the presence of $pp60^{v\text{-}src}$ and adhesion plaque disruption[16-18] although there are some exceptions to this. The $pp60^{v\text{-}src}$ proteins of the *ts* mutants NY68 and LA29 are both temperature-sensitive for adhesion plaque localization such that when these cells are grown at the restrictive temperature and are therefore phenotypically normal, $pp60^{v\text{-}src}$ is absent from adhesion structures.[16,19] Several mutants which induce a flat morphology or fusiform morphology also lack adhesion plaque-concentrated $pp60^{v\text{-}src}$.[16] However the RSV mutant CU2 confers a flat or blebby morphology but these cells display well-developed adhesion plaques which contain elevated levels of $pp60^{v\text{-}src}$. Other mutants such as CU11 and

Table 4.1. Cytoskeletal association and transformation parameters of RSV mutants and other viral tyrosine kinases

Virus	Morphol	Cell-Surface FN	CSK Association	Stress Fibers
None	N	+++	-	+++
w.t. RSV	R	-	++	-
tsNY68-35	R	-	++	-
tsNY68-41	N	+++	-	+++
tsLA29-35	R	-	+	-
tsLA29-41	N	+++	-	+++
tsLA32-35	R/F	-	+++	-
tsLA32-41	N	+++	++	+++
tsCUI I-35	N	-	+	-
tsCUI I-41	N	+	-	+
CU2	N	-	+++	++
RD10	N	-	++	+
CU12	F	+	-	-
tdSF104F	F	++	-	n.r.
dl15f	F	+++	-	n.r.
dl15r	R	-	++	n.r.
BSf	F	+++	-	n.r.
BSr	R	-	++	n.r.
Flf	F	++	-	n.r.
ffRSV	F	+++	n.r.	+
rASV2234	F	+ + +	++	+
FSV (fps)	F/R	+	++	+
PRCII (fps)	F	++	n.r.	+
AmuLV (abl)	R	-	+++	-
Y73 (yes)	F	+	++	+
AEV (ErbB)	F	+	++	+
UR2 (ros)	F	++	-	+

N, Normal, flat; R, rounded, transformed; F, fusiform; n.r., not recorded.
Modified from Kellie, S. BioEssays 1988; 8:25-30. Courtesy of company of Biologists.

RD10 exhibit a near normal morphology despite having pp60$^{v\text{-}src}$ clearly concentrated in the adhesion plaques. Furthermore our laboratory has examined cells transformed by other RSV mutants such as *ts*LA32 and rASV 2234.3 and found no strict correlation between pp60$^{v\text{-}src}$ in adhesion plaques and a rounded morphology (Fig. 4.1).[8,16,19-21] Thus the presence of pp60$^{v\text{-}src}$ in adhesion plaques is not sufficient to disrupt them. Many cells display a similar inverse relationship between pp60$^{v\text{-}src}$ in adhesion plaques and the loss of surface-associated fibronectin expression. However cells transformed by the fusiform mutants rASV 2234.3, 1702 and 157 all have prominent pp60$^{v\text{-}src}$-containing adhesion plaques plus a well-developed extracellular matrix. Finally *ts*LA32 cells grown at the restrictive temperature are phenotypically normal yet have apparently normal adhesion plaques containing kinase-active pp60$^{v\text{-}src}$. One explanation might be that although most of the mutant pp60$^{v\text{-}src}$ proteins are kinase active, they no longer recognize crucial adhesion plaque proteins as substrates although they can still correctly localize to adhesion plaques, and this will be discussed later. Alternatively since the studies described here only investigated the adhesion plaque localization of pp60$^{v\text{-}src}$ it is possible that the association of pp60$^{v\text{-}src}$ with other cytoskeletal components not necessarily located in adhesion plaques is essential for transformation.

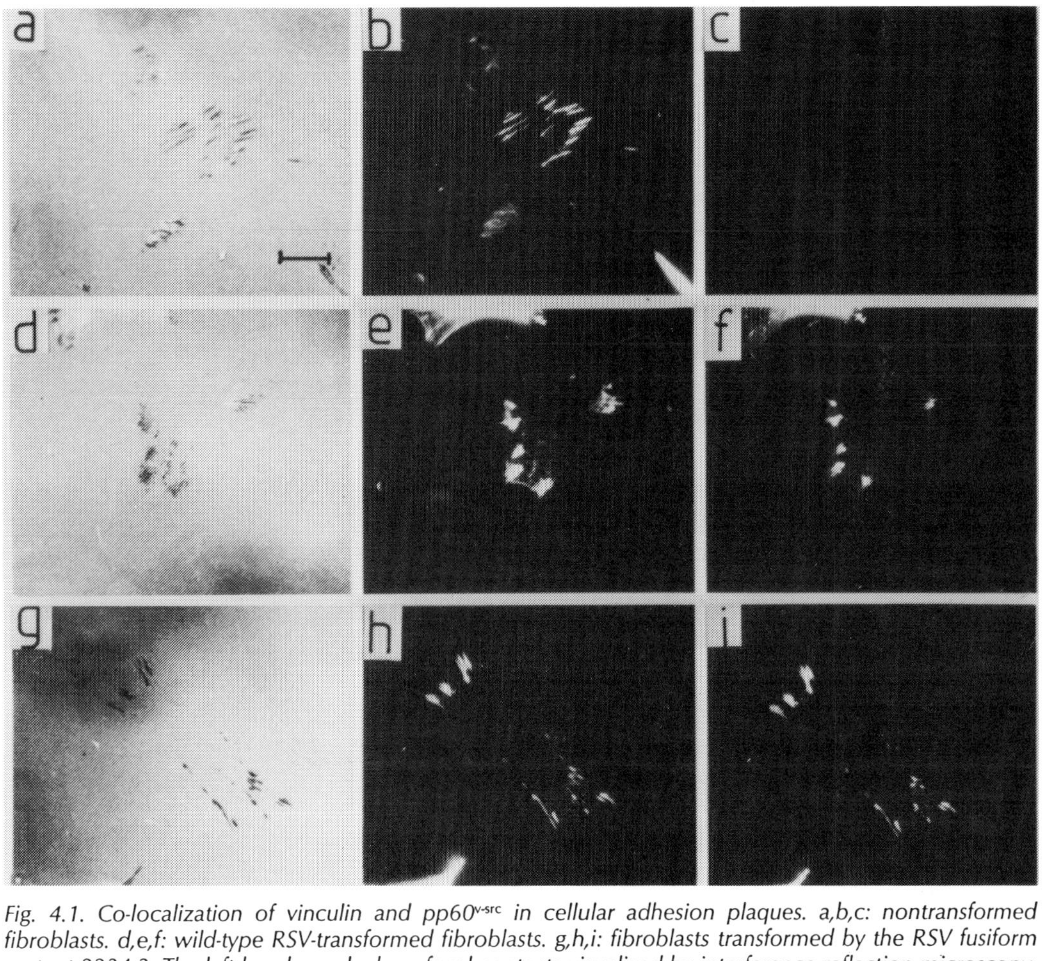

Fig. 4.1. Co-localization of vinculin and pp60$^{v\text{-}src}$ in cellular adhesion plaques. a,b,c: nontransformed fibroblasts. d,e,f: wild-type RSV-transformed fibroblasts. g,h,i: fibroblasts transformed by the RSV fusiform mutant 2234.3. The left hand panels show focal contacts visualized by interference reflection microscopy, the middle panels show vinculin localization and the right hand panls show pp60$^{v\text{-}src}$ localization. Figure courtesy of Kellie et al Exp Cell Res 1986; 165:216-228.

The Association of pp60$^{v\text{-}src}$ with the Cytoskeleton

The cytoskeleton is often defined as the filamentous residue left after extraction of cells with buffers containing nonionic detergents such as Triton X-100 or Nonidet P40. In an attempt to biochemically localize pp60$^{v\text{-}src}$ Burr et al[22] examined the Triton cytoskeletons of RSV-transformed fibroblasts for pp60$^{v\text{-}src}$ association and kinase activity. Virtually all of the pp60$^{v\text{-}src}$-associated kinase activity remained associated with the cytoskeletal fraction when the cells were extracted under these gentle conditions. Since more than 75% of the cellular proteins were extracted by this procedure a physical association of pp60$^{v\text{-}src}$ with the detergent-insoluble cytoskeleton is indicated.[22,23] Furthermore these workers found that pp60$^{v\text{-}src}$ is phosphorylated in situ when the cytoskeletal preparations were incubated with ^{32}P-γATP. Using antibodies against phosphotyrosine Comoglio et al[15] showed that most of the cellular tyrosine phosphorylated proteins were also associated with the cytoskeleton (see, for example Fig. 4.2). This physical association with the detergent-insoluble cytoskeleton, allied to the observations that cytoskeletal changes occur rapidly after pp60$^{v\text{-}src}$ is activated,[25] led to speculation that cytoskeletal proteins might be specific targets for pp60$^{v\text{-}src}$ and that tyrosine-specific phosphorylation of one or more of these proteins would lead to the cytoskeletal disruption which characterizes transformation.

Is the association of pp60$^{v\text{-}src}$ with the cytoskeleton essential for transformation? In support of this hypothesis Hamaguchi and Hanafusa[25,26] have reported such a correlation. To investigate the association of pp60$^{v\text{-}src}$ with the cytoskeleton in more detail Felice et al[23] used a combination of western blotting, kinase assays and immunofluorescence on cells transformed by the *ts* mutant LA29, whose properties were described earlier. At the permissive temperature (35°C) almost all the LA29 pp60$^{v\text{-}src}$ was associated with the cytoskeleton, and this protein was kinase active. Double immunofluorescent staining of these cells at 35°C showed that pp60$^{v\text{-}src}$ was concentrated in aggregates at the plane of the ventral membrane, many of which stained strongly for F-actin. These aggregates have a structure similar to those which have been previously described as podosomes or rosettes in other cell types. Extraction of these cells revealed that both the pp60$^{v\text{-}src}$ and actin in these podosomes were detergent insoluble and thus associated with cytoskeletal structures. At the restrictive temperature (39°C) however, LA29pp60$^{v\text{-}src}$ was mainly detergent-soluble, and the kinase activity of both LA29 detergent-soluble and insoluble material was

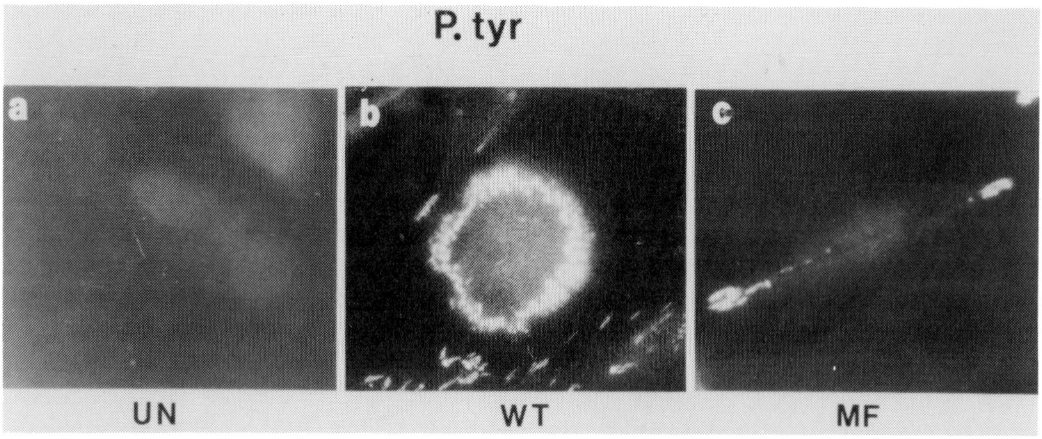

Fig. 4.2. Localization of tyrosine-phosphorylated proteins in normal and RSV-transformed chick embryo fibroblasts. Cells were labelled with PY20 monoclonal antibody against phosphotyrosine. a: nontransformed cells. b: Wild-type Prague C RSV-transformed cells. c: cells transformed by the RSV fusiform mutant 2234.3. Figure courtesy of the Company of Biologists: SEB Symposium 1992; 47: 267-282.

greatly reduced, indicating a defect in the *src* kinase at this temperature. Double immunofluorescence of these cells also revealed an increase in cytosolic pp60$^{v\text{-}src}$ as indicated by more diffuse staining, although a significant amount of pp60$^{v\text{-}src}$ remained associated with adhesion plaques coinciding with the termini of microfilament bundles. Therefore two distinct methods for detecting pp60$^{v\text{-}src}$ association with cytoskeletal structures, i.e., immunofluorescence and western blotting indicated that at 39°C LA29pp60$^{v\text{-}src}$ has a decreased ability to associate with the cytoskeleton.[22]

Similar experiments were performed to investigate the localization of phosphotyrosine-containing proteins. Western blotting using antiphosphotyrosine antibodies revealed several major phosphotyrosine-containing cytoskeletal-associated proteins in LA29-transformed cells cultured at 35°C, with the phosphotyrosyl protein content markedly reduced at 39°C, presumably due to the temperature sensitivity of LA29 pp60$^{v\text{-}src}$. At 35°C the anti-phosphotyrosine immunofluorescence in whole cells showed a podosomal localization which was resistant to detergent extraction. At the restrictive temperature phosphotyrosine staining was faint and was almost entirely eliminated by detergent extraction. The immunofluorescence experiments could not determine whether the phosphotyrosyl protein(s) present in adhesion plaques was pp60$^{v\text{-}src}$ itself, or whether other adhesion plaque proteins were also tyrosine phosphorylated, however since western blotting of the cytoskeletal fraction revealed several phosphotyrosine-containing proteins distinct from pp60$^{v\text{-}src}$, it seems likely that there are other phosphotyrosyl proteins present in the adhesion plaques of transformed cells. These potential substrates for pp60$^{v\text{-}src}$ will be discussed later. Since the kinase activities and cellular localization of pp60$^{v\text{-}src}$ were temperature-sensitive and correlated with transformation, this *ts* virus was used to study the role of the cytoskeletal association of pp60$^{v\text{-}src}$ during the early stages of transformation. When *ts*LA29-infected cells were grown at the restrictive temperature the pp60$^{v\text{-}src}$ was kinase inactive. If these cells were then switched to the permissive temperature there was a progressive increase in the kinase activity, detectable within 1-2 hours of temperature switch (Fig. 4.3). The localization of pp60$^{v\text{-}src}$ in the inactive and active states was investigated in these cells by fractionation into cytoskeletal or detergent-soluble phases followed by immune precipitation and pp60$^{v\text{-}src}$ kinase assays. The results indicated that concurrent with the temperature-induced kinase activation there was an initiation of association with the cytoskeleton such that all the detectable pp60$^{v\text{-}src}$-associated kinase activity occurred in the cytoskeletal fraction (Fig. 4.4). The mutation in *ts*LA29 occurs at the extreme carboxy terminus near the catalytic domain,[27] thus a single point mutation induces defects in both kinase activity and

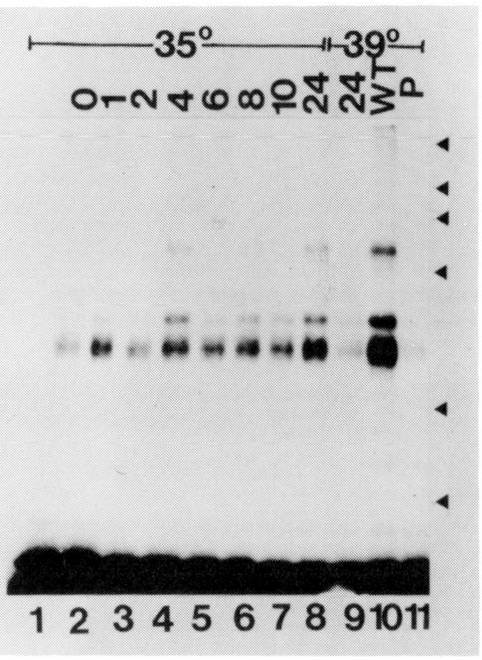

Fig. 4.3 . Time course of pp60$^{v\text{-}src}$-associated kinase activity in tsLA29-transformed cells. Cells were grown at 39°C for 48 hours then switched to 35°C. At the indicated times pp60$^{v\text{-}src}$ was immune precipitated and an in vitro kinase assay performed . WT: wild-type RSV-transformed cells. P: Pre-immune serum control. Numbers at the top are hours after temperature switch from 39°C to 35°C. Figure courtesy of the Company of Biologists: SEB Symposium 1992; 47: 267-282.

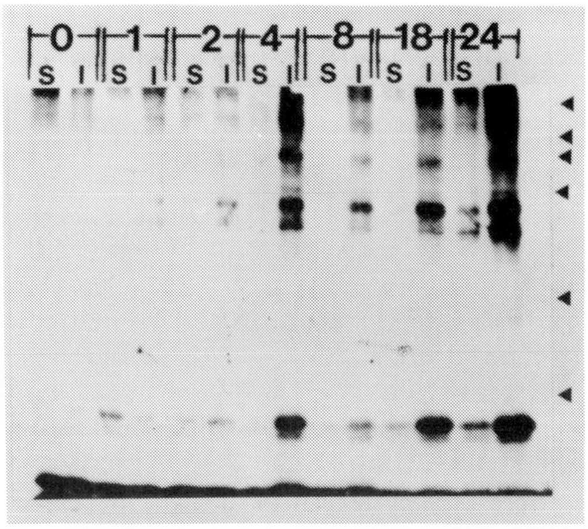

Fig. 4.4. Time course of association of kinase active pp60$^{v\text{-}src}$ with the cytoskeleton. Cells were grown as described for Fig. 2 and at the indicated times fractionated into cytoskeletal and detergent-soluble phases. pp60$^{v\text{-}src}$ was immune precipitated from each fraction and kinase assay performed. S: detergent-soluble phase. I: detergent-insoluble (cytoskeleton) phase. The pp60$^{v\text{-}src}$ kinase activity became associated with the cytoskeleton with a similar time course to enzymatic activation. Figure courtesy of the Company of Biologists: SEB Symposium 1992; 47: 267-282.

Table 4.2. Substrates for pp60$^{v\text{-}src}$ and their intracellular localizations

Substrate	Cellular Location
E-cadherin	intercellular junctions
β-catenin	intercellular junctions
connexin 43	intercellular junctions
vinculin	cell-substrate adhesions
talin	cell-substrate adhesions
integrin	cell-substrate adhesions
paxillin	cell-substrate adhesions
FAK	cell-substrate adhesions
annexin II	cytoskeleton
clathrin	coated pits
cortactin	cytoskeleton
ezrin	cytoskeleton
p110	cytoskeleton
tensin	cytoskeleton
calmodulin	cytoskeleton
caveolin	noncoated pits
enolase	glycolytic enzyme
phosphoglycerate mutase	glycolytic enzyme
lactate dehydrogenase	glycolytic enzyme
MAP kinase	cytoplasm/membrane
PLCγ	membrane
PI3 kinase	membrane
cdc2 kinase	cytoplasm
GAP	cytoplasm/membrane

cytoskeletal association and these deficiencies correlate with cellular transformation.

pp60$^{v\text{-}src}$ is not the only transforming tyrosine kinase which associates with the cytoskeleton. The transforming protein of murine leukemia virus is a p120$^{gag\text{-}abl}$ fusion protein with tyrosine kinase activity.[28] Immunofluorescence of this protein in transformed fibroblasts indicated that it was concentrated at the plasma membrane and cell-cell contacts, however the brightest fluorescence occurred at adhesion plaques and co-localized with vinculin.[28] Similarly the p80$^{gag\text{-}yes}$ transforming protein of Esh sarcoma virus and the p90$^{gag\text{-}yes}$ protein of Yamaguchi 73 sarcoma virus were detected in cell-substratum contacts.[29]

CYTOSKELETAL SUBSTRATES FOR PP60$^{V\text{-}SRC}$

The investigation of targets for pp60$^{v\text{-}src}$ within whole cells has proved complex, for example the comparison of phosphoproteins from normal and RSV-transformed cells by two-dimensional SDS-PAGE has yielded patterns which are difficult to interpret. Another approach has been to identify potential targets for pp60$^{v\text{-}src}$ phosphorylation within cells and to investigate these defined proteins. Since cytoskeletal changes occur rapidly after transformation, it has been suggested that cytoskeletal proteins are primary targets for pp60$^{v\text{-}src}$ (Table 4.2).

Several cytoskeletal proteins whose phosphorylation correlates with transformation have recently been described. Hamaguchi and Hanafusa[25,26] found that the phosphotyrosine-containing proteins were predominantly found in the Triton X-100-resistant fraction of cells transformed by Rous sarcoma virus, Fujinami sarcoma virus, PRCII or Y73, sub-fractionation revealing an association with crude membrane skeletons. These cytoskeletal pp60$^{v\text{-}src}$ substrates were characterized by Glenney and Zokas[30] who developed a panel of monoclonal antibodies to these proteins. Not only did these antibodies stain cytoskeletal structures, but antibodies against a 76 kDa protein and another against a 125 kDa protein stained adhesion plaques. The 125 kDa protein has now been characterized as focal adhesion kinase (FAK), a tyrosine kinase whose activity is deregulated in transformed cells and which will be described in more detail later.

Tyrosine Phosphorylation of Proteins in Intercellular Adhesion Structures

In addition to phosphorylating cytoskeletal proteins, pp60$^{v\text{-}src}$ induces tyrosine phosphorylation of components of both intercellular and extracellular matrix adhesion structures. As well as being concentrated in cell-substratum adhesions as already described, pp60^{src} is also concentrated in cell-cell adherens junctions. Utilizing an isolation system which preserved tyrosine phosphorylation by perfusing animals with the potent tyrosine phosphatase inhibitor vanadate before sacrifice, Tsukita et al[31] have shown that in normal adult rat liver cell-cell adherens junctions contain concentrations of phosphotyrosine-containing proteins. In addition these intercellular junctions also stained strongly for *c-src* and *c-yes* tyrosine kinases, and in vitro phosphorylation studies of isolated adherens junctions allowed biochemical identification of these kinases. Tyrosine phosphorylation of specific intercellular adherens junction proteins has also been detected. The stable association between cells is mediated by a family of calcium-dependent transmembrane adhesion molecules termed cadherins[32] which interact with the cytoskeleton via cytoplasmic molecules, the cadherin-associated proteins a and β catenins. After v-*src* transformation, cells have a reduced ability to adhere in a cadherin-dependent manner and the cadherin has been shown to be tyrosine phosphorylated.[33-35] Using temperature-sensitive *v-src* mutants both E-cadherin and β-catenin were shown to be tyrosine phosphorylated in epithelial cells and this phosphorylation correlated with the loss of intercellular adhesion. Treatment of transformed cells with the tyrosine kinase inhibitor herbimycin A reduced their tyrosine phosphorylation and increased cadherin-dependent adhesion.[36,37] In addition tyrosine phosphorylation of cadherins and catenins was not found in cells infected with nontransforming, nonmyristoylated *src* mutants.

The movement of small molecules between cells via gap junctions is inhibited by transformation by RSV. By utilizing temperature-sensitive mutants a rapid and reversible inhibition of intercellular communication has been demonstrated following a switch from the restrictive to the permissive temperature, indicating that expression of functional *v-src* protein is sufficient for inhibition of communication.[38,39] Inhibition of gap junction function also occurs if cells are transfected with activated *c-src* or by transformation by polyoma middle T antigen which is thought to mediate its effects through the activation of *c-src*.[40,41] Tyrosine phosphorylation of the gap junction protein connexin 43, with a concomitant reduction in gap junctional communication has also been reported.[42] The phosphorylation of connexin 43 is comparatively specific. When pp60$^{v\text{-}src}$ and connexin 43 or connexin 32 were co-transfected into oocyte pairs, pp60$^{v\text{-}src}$ completely abolished the communication induced by connexin 43, but had only a minor effect on communication induced by connexin 32. In these cells pp60$^{v\text{-}src}$ caused tyrosine phosphorylation of connexin 43 but not connexin 32. It seems likely that tyrosine-specific phosphorylation leads to inhibition of connexin 43-mediated intercellular communication by interfering with the gating of the channel or with multimeric assembly.[43] Thus proteins involved in cell-cell adhesion are good candidates for functional substrates of pp60$^{v\text{-}src}$. It remains to be demonstrated whether tyrosine phosphorylation of these proteins alters the protein-protein interaction of these molecules. The reduced intercellular adhesion between transformed cells might therefore be due to tyrosine phosphorylation of proteins which associate with intercellular junctions.

Tyrosine Phosphorylation in Cell-Matrix Adhesion Structures

Other defined proteins of adhesion structures which have been shown to be phosphorylated on tyrosine in RSV-transformed cells include vinculin, talin, paxillin and the transmembrane receptor for fibronectin, a member of the integrin superfamily of α/β heterodimers.[44-47] These proteins are found in cellular adhesion plaques, where pp60$^{v\text{-}src}$ is concentrated,[14] thus providing a physical co-distribution of potential substrates with enzyme within the cell. In association with other proteins, vinculin, talin and the fibronectin receptor are thought to form a transmembrane bridge linking the extracellular matrix with the intracellular actin-based cytoskeleton.[48-52] It is possible that phosphorylation of any one of these proteins might disrupt the adhesion plaque/cytoskeleton integrity resulting in the rounding up of the cells.

Vinculin

The discovery that vinculin from RSV-transformed cells contained 10-20 times as much phosphotyrosine as that of nontransformed cell suggested a model of adhesion plaque disruption as a consequence of vinculin phosphorylation.[44,53] As with many studies, mutant viruses which express variant pp60$^{v\text{-}src}$ proteins and confer different transformed phenotypes have been useful in examining whether tyrosine-specific phosphorylation of vinculin was responsible for the *src*-mediated cytoskeletal disruption and transformed morphology. Rohrschneider and Rosok examined a number of such mutants and found that many of those which conferred a flatter or fusiform morphology induced elevated tyrosine phosphorylation of vinculin, in many cases similar to wild-type.[16] However pp60$^{v\text{-}src}$ was also absent from the adhesion plaques of some of these mutants, indicating the possibility that these mutant pp60$^{v\text{-}src}$ proteins might preferentially phosphorylate a population of vinculin which is not associated with adhesion plaques. In other words, there may be a minor subpopulation of adhesion plaque-associated vinculin which is resistant to tyrosine phosphorylation and therefore can maintain the adhesion plaques of the fusiform cells. The role of phosphorylation in vinculin function has not yet been established, and so the relative importance of phosphorylated or nonphosphorylated vinculin in adhesion plaques is unknown. Kellie et al also studied a fusiform mutant where pp60$^{v\text{-}src}$ did localize in adhesion plaques and also found that vinculin from these cells was heavily tyrosine phosphorylated. In addition it was found that in cells infected by the *ts* mutant LA32 and grown at the restrictive temperature to give a normal phenotype the vinculin was tyrosine phosphorylated.[19] Cells which had reverted to a more normal morphology by the addition of exogenous fibronectin also retained an elevated level of phosphotyrosine on vinculin. Another example of lack of correlation between cell morphology and vinculin phosphorylation is the finding that vinculin phosphorylation does not change during the cellular rounding and loss of adhesion structures which accompanies normal cell division.[54-56] Thus tyrosine-specific phosphorylation of vinculin does not play a crucial role in the loss of adhesion plaque integrity induced by RSV transformation. The absence of a role for vinculin phosphorylation in transformation is further strengthened by investigations of other oncogenic tyrosine kinases. Although the *v-abl* and *v-yes* proteins induce tyrosine phosphorylation of vinculin the *v-ros* protein does not, and several viruses which encode *v-fps* proteins induce varying degrees of vinculin phosphorylation.[17,18,57] Stoichiometric calculations have shown that only 1-2% of vinculin molecules within transformed cells are tyrosine phosphorylated, i.e., 98% is nonphosphorylated. Therefore unless there are specific subpopulations of vinculin within the cell with specialized functions it is difficult to see how this low level of vinculin phosphorylation could lead to gross cytoskeletal changes. Nevertheless, the dissociation rate of vinculin from adhesion plaques of RSV-transformed cells is faster than the dissociation rate from nontransformed cells. It is possible that this might be a consequence of transformation-specific modifications of proteins to which vinculin binds rather than changes in vinculin itself.[58]

Talin

Using direct phospho-amino acid analysis or anti-phosphotyrosine antibodies, talin has been shown to be tyrosine phosphorylated in RSV-transformed cells.[45,46] Further analysis of *ts* mutants established that the phosphorylation was dependent on pp60$^{v\text{-}src}$ tyrosine kinase activity. However studies of the fusiform mutants which were described previously revealed that cells transformed by viruses which induced a flatter or fusiform morphology contained tyrosine-phosphorylated talin. Talin phosphorylation in cells transformed by other oncogenic tyrosine kinases was also inconsistent: transformation by Y73 induced talin phosphorylation but transformation by p130$^{v\text{-}fps}$ or *v-erb*B did not. Therefore there is a similar apparent lack of correlation between talin phosphorylation and cell rounding or cytoskeletal disruption as was found for vinculin.[47]

Integrins

Hirst et al[46] found that after transformation by RSV the β chain of the avian fibronectin receptor was phosphorylated on tyrosine. This phosphorylation was also observed in cells transformed by other tyrosine kinases such as those encoded by Fujinami sarcoma virus, Y73 or avian erythroblastosis virus. The tyrosine phosphorylation site of the integrin β chain was identified as tyrosine 778 in the cytoplasmic domain. This tyrosine phosphorylation of β integrins appears to reduce their avidity for both fibronectin and talin. In agreement with the tyrosine phosphorylation of other cytoskeletal proteins, tyrosine phosphorylation of β integrin in cells infected with *ts* mutants was transformation-dependent, leading these workers to suggest that the fibronectin receptor may be a substrate target for pp60$^{v\text{-}src}$, and that phosphorylation of the fibronectin receptor may play a role in reducing cell-matrix interactions thereby contributing to cytoskeletal disorganization. To determine the functional relevance of integrin phosphorylation Horvath et al investigated this in cells transformed with ASV2234.3, a mutant which transformed cells but which induces a flat, fusiform morphology but retains well-developed adhesion plaques and apparently normal intracellular distribution of pp60$^{v\text{-}src}$. The fibronectin receptor was tyrosine-phosphorylated in rounded RSV-transformed cells which lose surface fibronectin, but this phosphorylation was markedly reduced in RSV2234.3-transformed cells which retain an extensive fibronectin network (Fig. 4.5).[21] Furthermore, wild-type pp60$^{v\text{-}src}$ efficiently phosphorylated a synthetic peptide containing the putative integrin tyrosine phosphorylation site corresponding to tyrosine 778 whereas RSV2234.3 pp60$^{v\text{-}src}$ had a reduced capacity for phosphorylation of this peptide. The reason for this is unknown, however the most likely explanation is that RSV2234.3 pp60$^{v\text{-}src}$ had an altered substrate specificity, since there was no defect in its intracellular localization (Fig. 4.6). The findings that the cytoplasmic domain of β1 subunit contains the signal for focal contact localization was consistent with the possibility that phosphorylation of this site disrupts integrin function.[59] However substitution of tyrosine 788 with phenylalanine had no functional effect when this protein was transfected into either normal or transformed mouse 3T3 cells where it retained the ability to form normal heterodimers.[60] Since only about half the integrin receptors in these cells contain mutated subunit (due to the presence of endogenous normal integrin proteins) this may not be sufficient for complete disruption or deregulation. Nevertheless, the sequence of aminoacids 788 to 799, particularly tyrosine 788 and lysine 789 is important in focal contact localization and possibly talin binding.

One of the problems encountered through studying these cytoskeletal proteins has been the lack of a functional assay

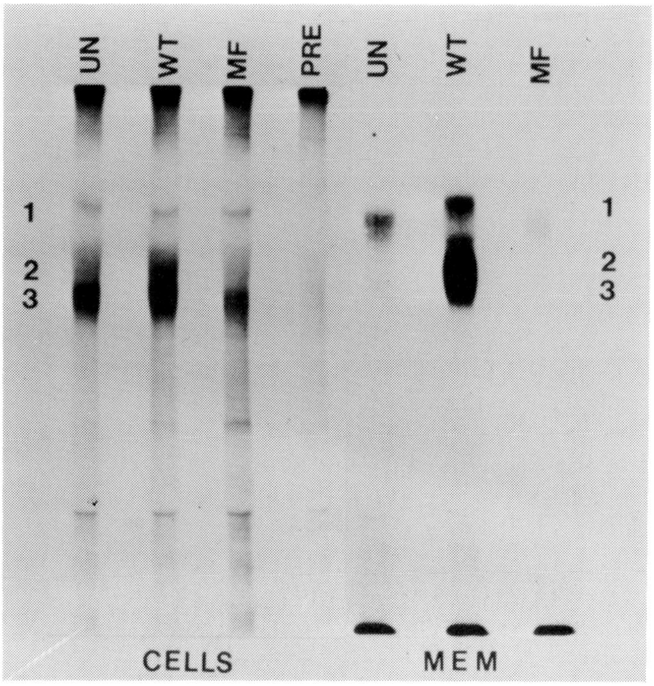

Fig. 4.5. Identification of tyrosine-phosphorylated CSAT integrin from nontransformed and RSV-transformed chick embryo fibroblasts by double immunoprecipitation. Autoradiographs of integrin immunoprecipitates from cells labelled with [^{32}P] orthophosphate (cells) or from in vitro kinase assays of cell membranes (MEM) in the presence of sodium orthovanadate. UN: nontransformed cells. WT: cells transformed by wild-type RSV. MF: cells transformed by RSV fusiform mutant 2234.3. Integrin from wild-type RSV-transformed cells is tyrosine phosphorylated while integrin from the fusiform RSV mutant has greatly reduced phosphotyrosine. 1,2,3 refer to CSAT bands 1,2 and 3 as described in Horvath et al.[21] Figure courtesy of Oncogene 1990; 5: 1349-1357, MacMillan Press.

within cells, although in vitro binding studies have been used successfully to demonstrate a physical interaction between vinculin and talin, and between talin and the fibronectin receptor.[50,61] Using such methods it has been possible to show that tyrosine-phosphorylated fibronectin receptor from RSV-transformed cells has a reduced capacity to interact with either talin or fibronectin.[62] In vitro α-actinin can interact with integrin β subunits therefore the phosphorylation of integrin might disrupt this association within cells, leading to adhesion plaque dissociation.[63] As is the case for vinculin and talin the stoichiometry of phosphorylation of integrins in transformed cells is low, and so unless there is a subpopulation of molecules containing high phosphorylation levels which preferentially localize to adhesion plaques it is difficult to explain the gross changes which occur in transformed cells purely on the basis of integrin phosphorylation. However such populations may indeed exist, as different integrin subpopulations with different phosphorylation profiles have been identified by the use of cell lysis in different detergent-containing buffers.[64]

Other explanations for integrin-related defects in transformed cells may exist. Plantefaber and Hynes[65] analyzed the expression of a variety of integrins in different cells following transformation and found that whilst levels of β1 and α3 remained unchanged, the expression of α5 was reduced and the expression of two other uncharacterized integrins, αA and αB was lost. In conclusion there is differential expression of high-affinity (α5) and low-affinity (α3) fibronectin receptors after transformation. The role of matrix receptors is further complicated, however, by the observation that immunohistochemical staining of integrin in RSV-induced tumors in chickens is more intense than that of the surrounding normal tissue, suggesting that the changes seen in vitro may not occur in vivo.

Paxillin

A novel adhesion plaque protein, paxillin, has recently been described which binds to the carboxyl terminal of vinculin.[66] This protein is phosphorylated on tyrosine in RSV-transformed cells with a stoichiometry of 20-30% (about 20-fold that of vinculin or talin) suggesting that it may have a functional consequence, although it remains to be demonstrated whether phosphorylation affects intermolecular association.[30,67] Interestingly, paxillin physically associates with pp60$^{v\text{-}src}$, therefore it is possible that the association of pp60$^{v\text{-}src}$ with focal contacts may be mediated by paxillin binding. Paxillin is also a major tyrosine kinase substrate in developing tissues.[68] Recently the region of *src* which binds to paxillin has been identified as the SH3 domain and it has been suggested that the SH3 domain of *src* may be a functional link to a pathway involved in cytoskeletal rearrangements.[69,70]

Integrin-mediated adhesion induces tyrosine phosphorylation of paxillin and focal adhesion kinase (FAK) in normal cells (as described below) and leads to a different

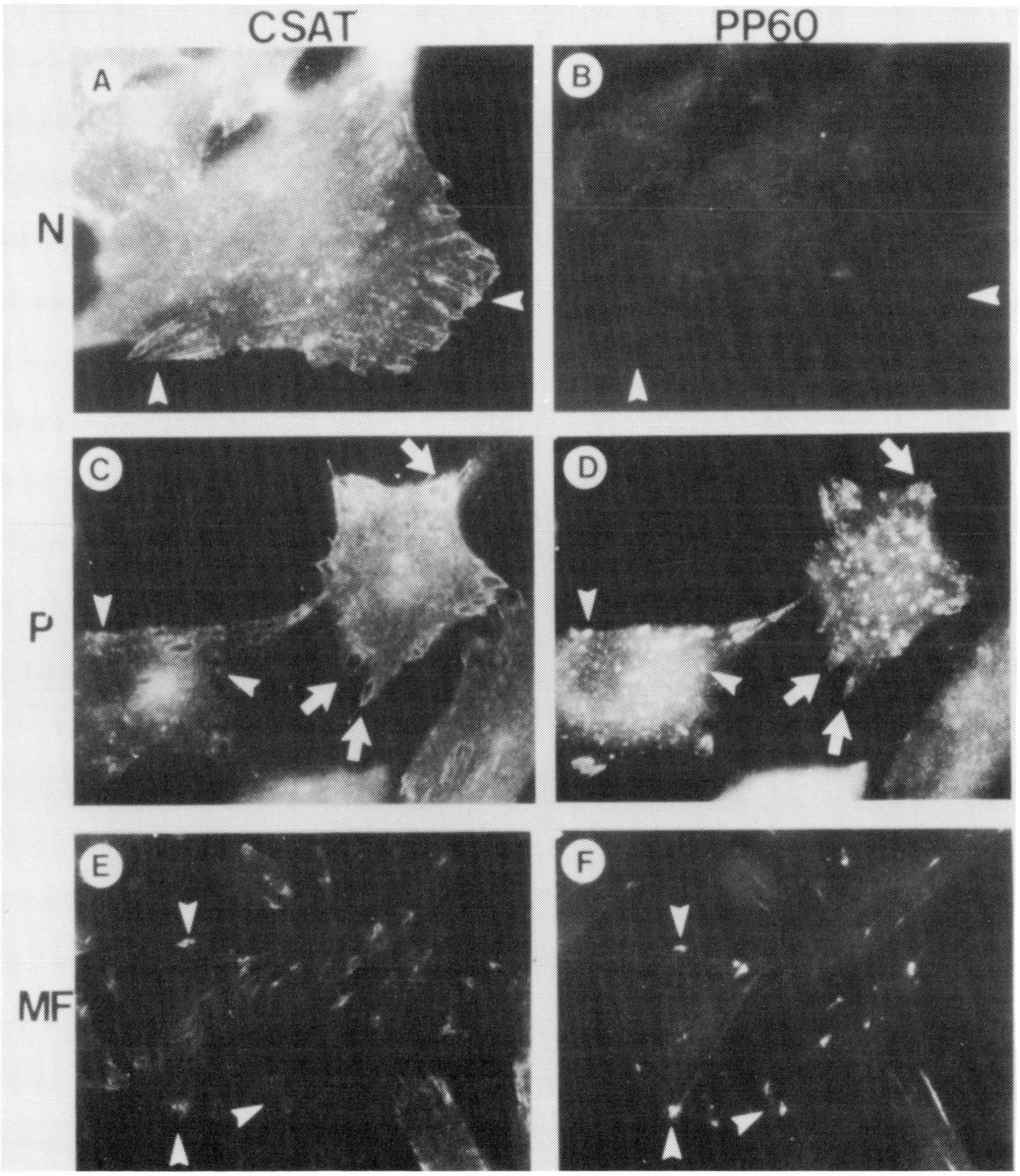

Fig. 4.6. Colocalization of integrin and pp60$^{v\text{-}src}$ in normal and RSV transformed chick embryo fibroblasts. Cells were double-labelled with anti-fibronectin receptor monoclonal antibody (left) and a rabbit antiserum against pp60$^{v\text{-}src}$ (right). A,B: Untransformed chick embryo fibroblasts. C,D: CEF transformed by wild-type RSV. E,F: CEF transformed by the fusiform mutant 2234.3. Bar = l0µm. Figure courtesy of Oncogene 1990; 5: 1349-1357, MacMillan Press.

set of responses from those in transformed cells: stress fiber formation instead of breakdown.[71] These events are linked since pretreatment of cells with a tyrosine kinase inhibitor such as herbimycin A abrogates the phosphorylation and stress fiber formation. One possible mechanism is that tyrosine-specific phosphorylation of paxillin might generate an SH2-binding domain which can be recognized by SH2-containing proteins such as tensin.[72] This is an actin crosslinking protein which might initiate actin nucleation and microfilament bundle formation, thus leading to assembly of the cytoskeleton. Why this should not happen in transformed cells is unclear but may relate to the uncoupling of cell adhesion from FAK activation as described below. The role of SH2 and SH3 domains in intracellular signaling and their importance in transformation will be discussed in the next chapter.

Focal Adhesion Kinase

Over the past few years there has been growing evidence of tyrosine kinase activity associated with the cytoskeleton of transformed (and nontransformed) cells which is independent of *c-src* or *v-src*. One of the proteins responsible for this has now been cloned, sequenced and characterized as a 125 kDa molecule called focal adhesion kinase ($pp125^{FAK}$). $pp125^{FAK}$ is unique as a tyrosine kinase since the catalytic domain is bordered by large amino-terminal and carboxy-terminal domains which do not contain SH2 or SH3 domains.[73] $pp125^{FAK}$ occurs in most tissues and a related 41kDa molecule, FAK-related nonkinase (FRNK) generated by alternative splicing of the FAK gene, comprising only the C-terminal noncatalytic domain has also been identified.[74] Immunofluorescence experiments have shown that both FAK and FRNK localize to adhesion plaques.[73,74] Crosslinking of a3β1 integrin by antibodies results in an elevation in the phosphotyrosine content of a 125 kDa identified as FAK.[75] Furthermore if fibroblasts are plated onto fibronectin, there is increased tyrosine phosphorylation of several proteins with molecular weights of 120-130 kDa, with the major tyrosine phosphorylated protein being $pp125^{FAK}$.[76] This does not occur when cells were plated on to poly-L-lysine or concanavalin A. Therefore it has been established that $pp125^{FAK}$ is itself a protein tyrosine kinase capable of autophosphorylation, and its activity is stimulated upon occupancy of the β_1 or β_3-containing fibronectin receptor, paralleling the increased phosphotyrosine content.[75,77-80] The activity of $pp125^{FAK}$ is probably important in the regulation of the cytoskeleton since within cells this protein co-localizes with other proteins such as paxillin. Paxillin also becomes phosphorylated on tyrosine in response to adhesion to fibronectin, and is a substrate for FAK in vitro in addition to being a substrate for $pp60^{v\text{-}src}$ as has been already described. Incubation of cells plated on to fibronectin in the presence of the tyrosine kinase inhibitor herbimycin A diminished the adhesion-induced tyrosine phosphorylation of both FAK and paxillin and inhibited the formation of focal contacts.[71,81] These results suggest that integrin-mediated tyrosine phosphorylation plays a role in the organization of the cytoskeleton. If this is so then it might be expected that this role would be affected by the profound cytoskeletal changes associated with transformation. $pp125^{FAK}$ was initially identified as a substrate for $pp60^{src}$ in *v-src*-transformed cells[82] and it displays transformation-enhanced enzymatic activity. To investigate whether FAK activity could be regulated by integrin occupancy and whether this was altered by oncogenic transformation Guan and Shalloway[78] studied the phosphorylation state and kinase activities of FAK in *src*-transformed cells plated onto fibronectin. Whilst in nontransformed cells $pp125^{FAK}$ tyrosine kinase activity was only induced after integrin occupancy, $pp125^{FAK}$ was constitutively active in *src*-transformed cells, although this activity could be further upregulated by integrin occupancy. The direct role of transformation in this deregulation was shown by the abrogation of this response in cells transfected with *src* mutants encoding nonmyristoylated or kinase negative pp60*src* proteins. In cells transfected with activated *src* or *fyn* FAK formed a complex with either pp60*src* or pp59*fyn*, which could be detected by co-immunoprecipitation from metabolically labeled cells, quantification revealing that more than 80% of the total cellular FAK associated with pp60*src*. Further studies with *src* and *fyn* mutants displaying deletions showed that FAK associated with their SH2 domains. This was confirmed by the findings that tyrosine-phosphorylated FAK bound to an isolated *src* SH2 domain in vitro. Interestingly, binding of FAK to the *src* SH2 domain rendered it resistant to phosphatase activity, indicating that this association may stabilize complex formation thereby enhancing FAK kinase activity by a positive feedback mechanism.[83] These results indicate that the effects of pp60*v-src* on the cytoskeleton may be indirect and could be mediated via upregulation of the tyrosine kinase activity of pp125FAK leading to hyperphosphorylation of cytoskeletal proteins. Cell adhesion mediated by integrins also results in the tyrosine-specific phosphorylation of paxillin by pp125FAK, and it is possible that the increased phosphorylation of paxillin found in transformed cells is due to a combination of pp60*src* and pp125FAK activities[84]. Other experiments which suggest a close interplay between FAK activity and cytoskeletal regulation come from the work of Rankin and Rozengurt[85] who have found that treatment of normal mouse fibroblasts with low doses (5ng/ml) of platelet-derived growth factor (PDGF) induces tyrosine phosphorylation of FAK and paxillin, whereas at higher concentrations (30ng/ml) there was no stimulation of phosphorylation. The low dose of PDGF induced actin stress fibers and membrane ruffles while at the high dose stress fibers were disrupted, indicating a correlation between actin organization and FAK phosphorylation. Furthermore cytochalasin D, which disrupts the actin cytoskeleton, completely inhibited the tyrosine phosphorylation of FAK and paxillin by PDGF, again indicating that phosphorylation of these focal contact proteins is critically dependent on the integrity of the actin cytoskeleton.

Tyrosine Phosphorylation of other Cytoskeletal Proteins

As I have already discussed, whilst several membrane and cytoskeletal targets for pp60*v-src* have now been identified, the functional significance of tyrosine-specific phosphorylation of these proteins is still debatable.[55] In vitro several cytoskeletal proteins have been shown to be substrates for pp60*v-src* including tubulin, MAP-2, fodrin and spectrin[86]. In addition, several other cytoskeletal proteins have been found to be phosphorylated on tyrosine in response to transformation in whole cells.

A Ca^{2+} and phospholipid-dependent actin-binding protein, p36 or annexin II,[87] is a major *src* substrate and as with other defined *src* substrates, it is a plasma membrane associated protein. Glenney[88] found that in vitro tyrosine phosphorylation of p36 reduced its ability to interact with purified actin, however the consequences of tyrosine phosphorylation within whole cells is unknown and suspect due to the low stoichiometry of phosphorylation. Investigations using partial transforming mutants revealed no correlation between the tyrosine-specific phosphorylation of p36 and any of the morphological events associated with transformation.[89] This was confirmed by the findings that p36 was constitutively phosphorylated in cells transformed by the *ts* mutant LA32 when these cells were grown at the phenotypically normal temperature.[8]

Clathrin, the major protein of coated vesicles, has been shown to be phosphorylated in RSV-transformed cells. Although not classed as an adhesion plaque protein, detailed electron microscopic studies have revealed that clathrin is often associated with the substrate-adherent ventral membranes.[90] Transformed cells also contain many dot-like focal contacts which are enriched in clathrin; therefore its phosphorylation may be a consequence of its close proximity to pp60*v-src*.[90] However, as with many of the cytoskeletal proteins described here there is no evidence that tyrosine phosphorylation of clathrin affects its function. Wu et al[91] described a novel substrate for pp60^{src} identified by isolation from a panel of monoclonal antibodies raised against phosphotyrosyl proteins. One of these antibodies specifically recognized two major protein species of 80 kDa and 85 kDa termed cortactin. Investigation of the intracellular localization of cortactin indicated a punctate cytoplasmic distribution with additional concentrations at cell extensions where it co-localized with F-actin. Further biochemical analysis revealed that cortactin was phosphorylated efficiently in cells transformed by mutants of *src* with defective or deleted SH2 or SH3 domains, which distinguishes it from other cytoskeletal substrates. Cloning of cortactin showed that it contained an SH3 domain and by analogy with other SH3 domain-containing proteins the SH3 domain of cortactin might be involved in protein-protein interactions. Deletion of the cortactin SH3 domain had no effect on its ability to bind actin, therefore the SH3 domain is not responsible for actin association, but might be important for its interaction with other cytoskeletal proteins.[92] Whether tyrosine phosphorylation affects actin binding or any other function of cortactin remains to be seen.

Ezrin (an 81 kDa protein originally found in brush border membranes), a plakoglobin-related protein, pp110 (a stress fiber-associated protein), tensin (a protein that caps and bundles F-actin), p22 caveolin (a protein found in clathrin-negative pits) are all phosphorylated on tyrosine after RSV transformation.[1,82,92-97] However, there is no data to indicate whether these are fortuitous phosphorylations due to the promiscuous substrate specificity of pp60$^{v\text{-}src}$, or whether they are responsible for the cellular changes associated with transformation. Tyrosine phosphorylation of calmodulin has also been detected after RSV transformation. Moreover when in vitro kinase assays using purified pp60$^{v\text{-}src}$ were performed the tyrosine phosphorylation of calmodulin was inhibited by calcium. The phosphorylated calmodulin had a decreased mobility in SDS-PAGE gels compared with unphosphorylated calmodulin in the presence of calcium, suggesting that the structure of the calmodulin-calcium complex had been altered by tyrosine phosphorylation. Fukami et al[93] suggested that calcium may regulate the level of tyrosine phosphorylation of calmodulin by pp60$^{v\text{-}src}$, and that this phosphorylation might attenuate the function of this protein in vivo. Further work is necessary to test this hypothesis.

THE EFFECT OF TRANSFORMATION ON THE PROTEINS OF MEMBRANE-CYTOSKELETAL STRUCTURES

If tyrosine-specific phosphorylation of membrane-cytoskeletal proteins is involved in maintenance or induction of the transformed phenotype, how could this lead to morphological changes? We can postulate at least two possible extremes of mechanisms for cell rounding. Tyrosine phosphorylation of the majority of adhesion plaque proteins results in a decreased intermolecular association of all these proteins, leading to a total break-up of adhesion structures and thence to cell rounding. An alternative explanation would propose that tyrosine phosphorylation of only one component, such as an integrin results in decreased cellular affinity for fibronectin, leading to cell rounding, with the break-up of adhesion structures as a secondary process, i.e., adhesion plaque disruption is an consequence of cell rounding, not a cause (Fig. 4.7). Vinculin has been found in the residual focal contacts, podosomes and rosettes of transformed cells,[18,19,23,54,98-102] and co-localises with actin in these structures. Talin is also localized in these adherens structures and therefore the molecular composition of adhesion structures in transformed cells is similar to that of nontransformed cells, despite the ultrastructural changes. In addition clathrin has been found associated with point contacts of transformed cells.[90] These observations indicate that although some of the proteins have been lost, the ability of vinculin to interact with talin and/or adhesion structures is at least partly retained after transformation. The situation is probably more complex than this, since vinculin appears to be lost more rapidly than talin from adhesion plaques following transformation.[58] Thus there may be sequential disruption of intermolecular associations leading to a sequential loss of individual adhesion plaque proteins.

In nontransformed cells the fibronectin receptor is concentrated in adhesion plaques where it adopts an "eye of the needle" con-

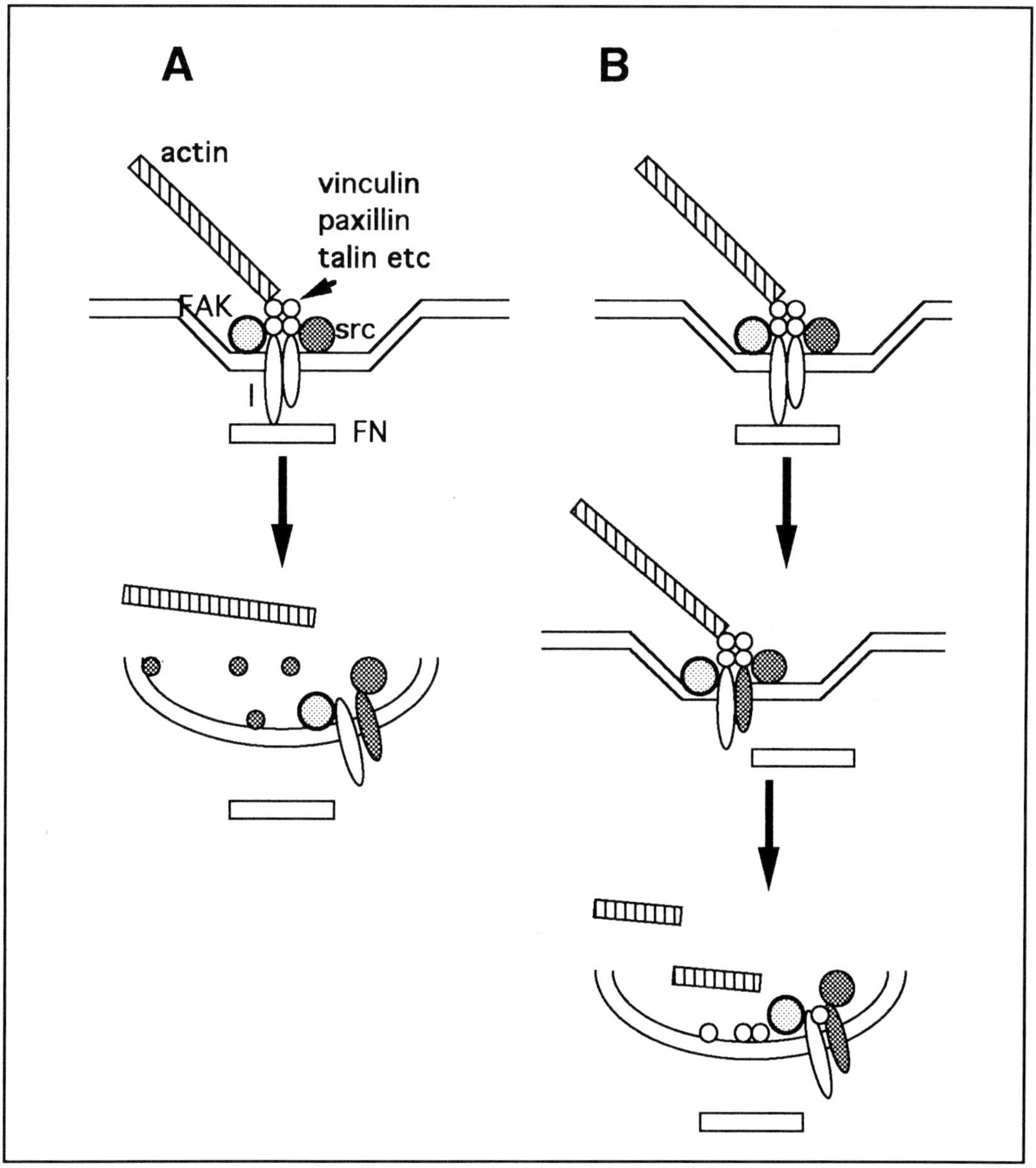

Fig. 4.7. Cartoon of two extreme consequences of v-src-induced tyrosine phosphorylation of adhesion plaque components. A: tyrosine phosphorylation of one or more adhesion plaque component leads to complete disruption of intermolecular associations and therefore adhesion plaque disruption and cell rounding. B: tyrosine phosphorylation of one component, e.g., the fibronectin receptor leads to breakdown of the interaction of fibronectin with its receptor, leading to cell rounding. Partial disruption of adhesion plaque associations is a consequence of this cell rounding. Figure courtesy of JAI Press, Treatise on the Cytoskeleton vol 4.

figuration and co-localizes with fibronectin fibrils. Nevertheless studies of RSV-transformed cells have revealed that it is more randomly distributed throughout the plasma membrane with little or no adhesion plaque localization.[21,103,104,105] This observation is consistent with the suggestion that phosphorylation results in a decreased interaction of the fibronectin receptor with fibronectin.[62] Since talin is not redistributed following transformation this implies that (a) the interaction of the fibronectin receptor with talin is also disrupted, and (b) the fibronectin receptor is not responsible for anchoring talin to adhesion structures. Therefore, whilst it is true that most adhesion plaques are disrupted upon transformation, some colocalization of actin-vinculin-talin-integrin-pp60$^{v\text{-}src}$ in the remaining adhesion structures is preserved, along with increased tyrosine phosphorylation within these structures.[15,21,23] Other investigations have also suggested that functional cytoskeletal interactions are not completely disrupted after transformation. The co-capping of talin with integrin was unaffected by transformation with RSV[106] and no differences were found in the ability of normal and transformed cells to cluster talin and actin in response to fibronectin-coated beads.[107] These reports are consistent with talin being more tightly associated with adhesion plaques than vinculin in transformed cells.[58] These results suggest that a consequence of tyrosine phosphorylation is the disruption of the interaction between the fibronectin receptor and fibronectin without necessarily affecting talin-vinculin or talin-vinculin-adhesion plaque interactions.

Whilst tyrosine phosphorylation of vinculin, talin and some other proteins has proved to be irrelevant for neoplastic transformation, these components are still important regulatory molecules in normal cell function. Overexpression of vinculin in fibroblasts suppresses cell motility, and a recent study has shown that transfection of vinculin cDNA into transformed cells which express low levels of vinculin results in suppression of tumorigenicity following injection into syngeneic hosts or nude mice.[108,109] Similarly, overexpression of the fibronectin receptor in CHO cells suppresses the transformed phenotype.[110]

OTHER POSSIBLE MECHANISMS FOR CYTOSKELETAL REARRANGEMENTS

It is entirely possible that the changes associated with the cytoskeleton of transformed cells may occur indirectly as a consequence of tyrosine phosphorylation of proteins which are not considered to be cytoskeletal components or by aberrant gene expression of cytoskeletal proteins. One example of the former might be changes in the activity of the actin regulating protein profilin due to tyrosine phosphorylation and subsequent activation of phospholipase Cγ (PLCγ). Profilin is a small protein (12-15 kDa) which binds to monomeric actin in vitro thereby preventing the formation of polymeric actin filaments by sequestering the actin monomers when present at high concentrations.[111] However at the lower more physiological concentrations found within cells profilin can also promote actin polymerization and filament stabilization by increasing the net concentration of actin-ATP, which can polymerise much more rapidly than actin-ADP. Phosphatidylinositol-4,5-bisphosphate (PIP_2) binds to profilin with a stoichiometry of 4-5:1 and competes for actin binding, resulting in a dissociation of the profilin-actin complex.[112] Much of the profilin within the cell is localized to the intracellular face of the plasma membrane where it is likely to encounter high concentrations of PIP_2. Normally the binding of PIP_2 to profilin also protects this lipid from hydrolysis by PLCγ, however if PLCγ is tyrosine phosphorylated following growth factor stimulation or oncogenic transformation, this enzyme can overcome the profilin-mediated protection of PIP_2 and efficiently hydrolyse the lipid to generate inositol phosphate and diacylglycerol.[113] This would result in the release of profilin, potentially leading to increased actin nucleation and a change in the cellular actin equilibrium towards polymerization and filament stabilization.[114-118] This proposal appears to be the opposite of what might be expected, since there is a reduction in the number of stress fibers after transformation. However this increase in profilin/actin nucleation might explain the increased membrane ruffling and pseudopodial activity observed in transformed cells, and, in conjunction with breakdown of adhesion structures as described below, contribute to the enhanced motility of transformed cells.

Transformation also commonly results in enhanced diacylglycerol and calcium mobilization due to the activity of PLCγ which in turn activate protein kinase C (PKC). Both PKC and calcium can alter cytoskeletal integrity as a consequence of their actions upon different actin-binding proteins. For example the bundling activity of α-actinin is dependent upon calcium, therefore any change in the intracellular calcium concentration could lead to the reorganization of actin bundles.[119] One prime candidate for a protein whose function is altered in transformed cells is MARCKS (myristoylated, alanine rich, C kinase substrate). MARCKS normally associates with the intracellular face of the plasma membrane and can crosslink filamentous actin.[120] When PKC is activated MARCKS becomes phosphorylated with the result that MARCKS is released from the plasma membrane and loses its ability to promote actin crosslinking. Increases in intracellular calcium also promotes the binding of calmodulin to MARCKS, which also results in decreased actin crosslinking.[121] There has been a report that MARCKS is downregulated in transformed cells, providing a correlation between lowered levels of MARCKS, a reduction in microfilament bundling and cell rounding.[122]

INTRACELLULAR ASSOCIATIONS AND SUBSTRATES FOR PP60^{C-SRC}: RELATIONSHIPS WITH THE CYTOSKELETON

In comparison with the *v-src* protein, relatively little is known about the physiological role of its normal cellular homolog, pp60$^{c\text{-}src}$. Only two cell types express the cellular *src* gene product at readily detectable levels: neurons and blood platelets.[123-125] In neurons pp60$^{c\text{-}src}$ is concentrated in the growth cone, leading to speculation that tyrosine phosphorylation might be involved in neurite extension.[126] α and β tubulin and the synaptic vesicle protein synaptophasin have been shown to be substrates for pp60$^{c\text{-}src}$, and so it may be involved in neurosecretion in these cells.[127-130] The effect of tyrosine phosphorylation on the functions of these proteins has yet to be examined.

The intracellular localization of pp60$^{c\text{-}src}$ in fibroblasts which overexpress this protein following transfection of the *c-src* gene has shown that significant quantities associate with intracellular membranes in addition to the plasma membrane,[131-133] although in human blood platelets pp60$^{c\text{-}src}$ is restricted to the surface-associated membrane.[134,135] It is possible that this differential expression may be related to different cell functions. For example platelets are non-dividing cells whose primary responses to stimulation are increased adhesiveness and secretion. Fibroblasts have the capacity to proliferate and will respond to a number of stimuli by mitosis. Thus the interaction of pp60$^{c\text{-}src}$ with intracellular membranes may be important for the regulation of cell division. This proposal is consistent with the demonstration of differential association of pp60$^{c\text{-}src}$ with different intracellular membrane systems during the cell cycle.[131] The surface membrane association of pp60$^{c\text{-}src}$ in platelets may be more important for intracellular signaling processes unrelated to proliferation, such as modulation of adhesion molecules.[136,137]

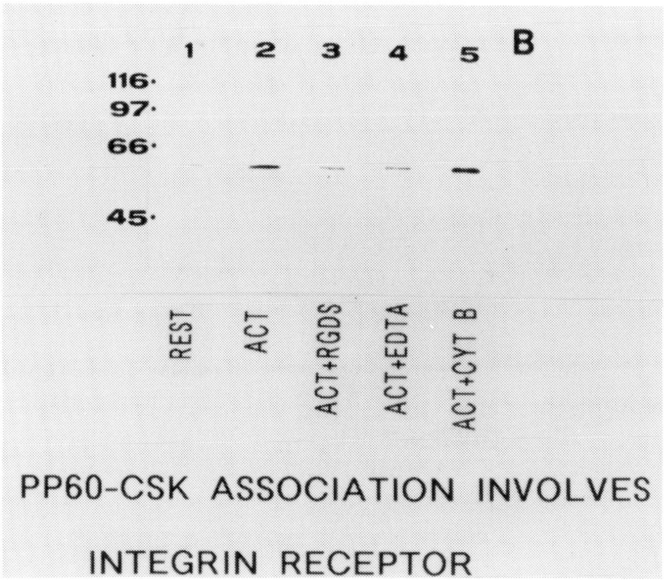

Fig. 4.8. Association of pp60^c-src^ with the cytoskeleton during different conditions of cell activation. Cytoskeletons from resting cells (1) or from platelets activated by 1U/ml thrombin (2) in the presence of 200 µg/ml RGDS peptide (3), or 10mM EDTA (4) or 10 µg/ml cytochalasin B (5). The association of pp60^c-src^ with the platelet cytoskeleton correlates with integrin receptor occupancy. Figure courtesy of EMBO J 1992; 11: 855-861.

Upon platelet activation $pp60^{c\text{-}src}$ translocates from a detergent-soluble to a detergent-insoluble phase.[137] Thus following activation $pp60^{c\text{-}src}$ partially converts from a membrane associated, non-cytoskeletal component to a membrane-associated cytoskeletal component where it complexes with gpIIb/IIIa (Fig. 4.8).[138] It is unclear at present whether these differentially located *c-src* proteins have differing functions. A minor subpopulation of $pp60^{c\text{-}src}$ associated with the membrane skeleton, rather than the cytoskeleton has also been detected in platelets, however this was unchanged after activation.[137]

Little is known about the substrates for $pp60^{c\text{-}src}$ within platelets, however two recent studies have identified the platelet fibrinogen receptor glycoprotein IIb/IIIa as a potential substrate.[136,139] This molecule is a transmembrane glycoprotein comprising two subunits, gpIIb and gpIIIa which can also bind fibronectin, laminin and thromospondin and belongs to the integrin superfamily of adhesion receptors. Findik et al[139] reported that purified platelet $pp60^{c\text{-}src}$ can phosphorylate both subunits of glycoprotein IIb/IIIa, however Elmore et al[136] demonstrated that in platelet membranes only the IIIa chain of gpIIb/IIIa is phosphorylated on tyrosine. As discussed previously immunopurified platelet $pp60^{c\text{-}src}$ can also phosphorylate a peptide containing the tyrosine phosphorylation site of fibroblast integrin β_1 subunit, which is partially conserved in the IIIa chain. Since thrombin-induced tyrosine-specific phosphorylation can be inhibited by RGDS peptides,[140] this implies a close functional association of $pp60^{c\text{-}src}$ with adhesion receptors during platelet activation. As described earlier FAK is activated when platelets are stimulated, and it is possible that the increased tyrosine phosphorylation is due to FAK activity.[64,79] The findings that integrins can act as substrates for both the transforming and non-transforming *src* proteins, with consequences which include changes in cell-matrix interactions and cytoskeletal reorganization strongly suggest that tyrosine-specific phosphorylation of integrin receptors plays a crucial role in the regulation of cell shape.

References

1. Eckhart W. Oncogenes of DNA tumor viruses: papavoviruses. In: Weinberg RA, ed. Oncogenes and the Molecular Origins of Cancer. New York: Cold Spring Harbor Laboratory Press, 1989: 223-238.
2. Bohmann D, Bos TJ, Admon A et al. Human protooncogene *c-jun* encodes a DNA-binding protein with structural and functional properties of a transcription factor AP-1. Science 1987; 238: 1386-1392.
3. Franza BR, Rauscher FJ, Josephs SF et al. The *fos* complex and *fos*-related antigens recognize sequence elements that contain AP-1 binding sites. Science 1988; 239:

1150-1153.

4. Rousell MF, Downing JR, Rettenmeier CW et al. A point mutation in the extracellular domain of the human CSF-1 receptor (c-*fms* protooncogene product) activates its transforming potential. Cell 1988; 55: 979-988.
5. Foster DA, Shibuya M, Hanafusa H. Activation of the transforming potential of the cellular *fps* gene. Cell 1985; 42: 105-115.
6. Willingham MC, Jay G, Pastan I. Localization of the ASV src gene product to the plasma membrane of transformed cells by elecron microscopic immunocytochemistry. Cell 1979; 18: 125-134
7. Kreuger JG, Wang E, Goldberg AR. Evidence that the *src* gene product is membrane associated.Virol 1980; 101: 25-40.
8. Stoker A, Kellie S, Wyke JA. Intracellular localization and processing of pp60$^{v\text{-}src}$ proteins expressed by two distinct temperature-sensitive mutants of Rous sarcoma virus. J Virol 1986; 58: 876-883.
9. Cross FR, Garber EA, Pellman D, Hanafusa H. A short sequence in the p60^{src} N terminus is required for p60^{src} myristylation and membrane association and for cell transformation. Mol Cell Biol 1984; 4: 1834-1842.
10. Kamps MP, Buss JE, Sefton BM. Rous sarcoma virus transforming protein lacking myristic acid phosphorylates known peptide substrates without inducing transformation. Cell 1986; 45: 105-112.
11. Wyke JA, Stoker A. Genetic analysis of the form and function of the viral *src* gene. Biochim Biophys Acta 1987; 907: 47-70.
12. Kaplan JM, Varmus HE, Bishop JM. The *src* protein contains multiple domains for specific attachment to membranes. Mol Cell Biol 1990; 10: 1000-1009.
13. Resh MD, Ling H-P. Identification of 32K a plasma membrane protein that binds to the myristylated amino-terminal sequence of pp60$^{v\text{-}src}$. Nature (London) 1990; 346: 84-86.
14. Rohrschneider LR. Adhesion plaques of Rous sarcoma virus-transformed cells contain the *src* gene product. Proc Natl Acad Sci USA 1980; 77: 3514-3518.
15. Comoglio PM, Direnzo MF, Tarone G et al. Detection of phosphotyrosine-containing proteins in the detergent-insoluble fraction of RSV-transformed fibroblasts by azobenzophosphanate antibodies. EMBO J 1984; 3: 483-489.
16. Rohrschneider L, Rosok MJ. Transformation parameters and pp60$^{v\text{-}src}$ localization in cells infected with partial transformation mutants of Rous sarcoma virus. Mol Cell Biol 1983; 3: 731-746.
17. Rohrschneider L, Reynolds S. Regulation of cellular morphology by Rous sarcoma virus *src* gene: analysis of fusiform mutants. Mol Cell Biol 1985; 5: 3097-3107.
18. Nigg EA, Sefton BA, Hunter T et al. Immunofluorescent localization of the transforming protein of Rous sarcoma virus with antibodies against a synthetic *src* peptide. Proc Natl Acad Sci USA 1986; 79: 5939-5942.
19. Kellie S, Patel B, Wigglesworth NM et al. The use of Rous sarcoma virus transformation mutants with differing tyrosine kinase activities to study the relationships between vinculin phosphorylation, pp60$^{v\text{-}src}$ location and adhesion plaque integrity. Exp Cell Res 1986; 165: 216-228.
20. Kreuger JG, Garber EA, Chin SS et al. Size-variant pp60^{src} proteins of recovered avian sarcoma viruses interact with adhesion plaques as peripheral membrane proteins: effect on cell transformation. Mol Cell Biol 1983; 4: 454-467.
21. Horvath AR, Elmore MA, Kellie S. Differential tyrosine-specific phosphorylation of integrin in Rous sarcoma virus transformed cells with differing transformed phenotype. Oncogene 1990; 5: 1349-1357.
22. Burr JG, Dreyfuss G, Penman S et al. Association of the *src* gene product of Rous sarcoma virus with cytoskeletal structures of chick embryo fibroblasts. Proc Natl Acad Sci USA 1981; 77: 3484-3488.
23. Felice GR, Eason P, Nermut MV, Kellie S. pp60$^{v\text{-}src}$ association with the cytoskeleton induces actin reorganization without affecting polymerization status. Eur J Cell Biol 1990; 52: 47-59.
24. Boschek CB, Jockusch BM, Friis RR et al. Early changes in the distribution and organization of micro-filament proteins during cell transformation. Cell 1981; 24: 175-184.
25. Hamaguchi M, Hanafusa H. Association of

pp60^{src} with triton X-100-resistant cellular structure correlates with morphological transformation. Proc Natl Acad Sci USA 1987; 84: 2312-2316.
26. Hamaguchi M, Hanafusa H. Localization of major potential substrates of p60$^{v\text{-}src}$ kinase in the plasma membrane matrix fraction. Oncogene Res. 1989; 4: 29-37.
27. Welham M, Wyke JA. Mitogenesis induced by pp60$^{v\text{-}src}$ is not accompanied by increased expression of immediate early response genes. Oncogene. 1990; 5: 161-169
28. Rohrschneider L and Najita LM. Detection of the *v-abl* gene product at cell-substratum contact sites in Abelson leukemia virus transformed cells. J Virol 1984; 51: 547-552.
29. Gentry LE and Rohrschneider LR. Common features of the *yes* and *src* gene products defined by peptide-specific antibodies. J Virol 1984; 52: 539-546.
30. Glenney JR, Zokas L. Novel tyrosine kinase substrates from Rous sarcoma virus transformed cells are present in the membrane skeleton. J Cell Biol 1989; 108: 2401-2408.
31. Tsukita S, Oishi K, Akiyama T et al. Specific proto-oncogenic tyrosine kinases of src family are enriched in cell-to-cell adherens junctions where the level of tyrosine phosphorylation is elevated. J Cell Biol 1991; 113: 867-879.
32. Tacheichi M. Cadherin cell adhesion receptors as a morphogenetic regulator. Science 1991; 251: 1451-1455.
33. Volberg T, Geiger B, Dror R et al. Modulation of intercellular adherens-type junctions and tyrosine phosphrylation of their components in RSV-transformed chick lens cells. Cell Reg. 1991; 2: 105-120.
34. Volberg T, Zick Y, Dror R et al. The effect of tyrosine-specific phosphorylation on the assembly of adherens-type junctions. EMBO J 1992; 11: 1733-1742.
35. Matsuyoshi N, Hamaguchi M, Taniguchi S .et al. Cadherin-mediated cell-cell adhesion is perturbed by v-*src* tyrosine phosphorylation in metastatic fibroblasts. J Cell Biol 1992; 118: 703-714.
36. Hamaguchi M, Matsuyoshi N, Ohnishi Y et al. p60$^{v\text{-}src}$ causes tyrosine phosphorylation and inactivation of the N-cadherin-catenin cell adhesion system. EMBO J 1993; 21: 307-314.
37. Behrens J, Vakaet L, Friis R et al. Loss of epithelial differentiation and gain of invasiveness correlates with tyrosine phosphorylation of the E-cadherin/β-catenin complex in cell transformed with a temperature-sensitive v-SRC gene. J Cell Biol 1993; 120: 757-766.
38. Atkinson MM, Menko AS, Johnson RG et al. Rapid and reversible reduction of junctional permeability in cell infected with a temperature sensitive mutant of avian sarcoma virus. J Cell Biol 1981; 91: 573-578.
39. Azarnia R, Lowenstein WR. Intercellular communication and the control of growth. X. Alteration of junctional permeability by thye src gene. Astudy with temperature sensitive mutant Rous sarcoma virus. J Membr Biol 1984; 82: 191-205.
40. Azarnia R, Reddy S, Kmiecik TE et al. The cellular *src* gene product regulates junctional cell-to-cell communication. Science 1988; 239: 398-401.
41. Azarnia R, Lowenstein WR. Polyoma virus middle-T antigen down regulates junctional cell-cell communication. Mol Cell Biol 1987; 7: 946-950.
42. Crow DS, Beyer EC, Paul DL et al. Phosphorylation of connexin-43 gap junction protein in uninfected and RSV-transformed mammalian fibroblasts. Mol Cell Biol 1990; 10: 1754-1780.
43. Swenson KL, Piwnica-Worms H, McNamee H, Paul DL. Tyrosine phosphorylation of the gap junctional protein connexin-43 is required for the pp60$^{v\text{-}src}$ -induced inhibition of communication. Cell Regulation 1990; 1: 989-1002.
44. Sefton BM, Hunter T, Ball EH, & Singer SJ. Vinculin: A cytoskeletal target of the transforming protein of Rous sarcoma virus. Cell 1981; 24: 165-174.
45. Pasquale EB, Maher PA, Singer SJ. Talin is phosphorylated on tyrosine in chicken embryo fibroblasts transformed by Rous sarcoma virus. Proc Natl Acad Sci USA 1986; 83: 5507-5511.
46. Hirst R, Horwitz A, Buck C, Rohrshneider L. Phosphorylation of the fibronectin recep-

tor complex in cells transformed by oncogenes that encode tyrosine kinases. Proc Natl Acad Sci USA 1986; 83: 6470-6474.

47. DeClue JE, Martin GS. Phosphorylation of talin at tyrosine in Rous sarcoma virus transformed cells. Mol Cell Biol 1987; 7: 371-378.
48. Burridge K. Substrate adhesions in normal and transformed fibroblasts: organization and regulation of cytoskeletal, membrane and extracellular matrix components at focal contacts. Cancer Rev 1986; 4: 18-78.
49. Burridge K, Fath K, Kelly T et al. Focal adhesions: transmembrane junctions between extracellular matrix and the cytoskeleton. Ann Rev Cell Biol 1988; 4: 487-525.
50. Luna EL, Hitt AL. Cytoskeleton-plasma membrane interactions. Science 1992; 258: 955-964.
51. Jones P, Jackson P, Price GJ et al. Identification of a talin binding site in the cytoskeletal protein vinculin. J Cell Biol 1989; 109: 1917-2927.
52. Nuckolls GH, Turner CE, Burridge K. Functional studies of the domains of talin. J Cell Biol 1990; 110, 1635-1644.
53. Hynes RO. Tyrosine phosphorylation of vinculin by pp60$^{v\text{-}src}$: what might it mean? Cell 1982; 28: 437-438.
54. Kellie S. Cellular transformation, tyrosine kinase oncogenes and the cellular adhesion plaque. BioEssays 1988; 8: 25-30.
55. Kellie S, Patel B, Wigglesworth NM et al. Comparison of the relative importance of tyrosine-specific vinculin phosphorylation and the loss of surface-associated fibronectin in the morphology of cells transformed by Rous sarcoma virus. J Cell Sci 1986; 82: 129-142.
56. Rosok MJ, Rohrschneider L. Increased phosphorylation of vinculin on tyrosine does not occur during the release of stress fibers before mitosis in normal cells. Mol Cell Biol 1983; 3: 475-479.
57. Antler AM, Greenberg ME, Edelman GM et al. Increased phosphorylation of tyrosine in vinculin does not occur upon transformation by some avian sarcoma viruses. Mol Cell Biol 1985; 5: 263-267.
58. Brands R, De Boer A, Feltkamp CA et al. Disintegration of adhesion plaques in chick embryo fibroblasts upon Rous sarcoma virus transformation: different dissociation rates for talin and vinculin. Exptl Cell Res 1990; 186: 138-148.
59. Solowska J, Guan J-L, Marcantonio EE et al. Expression of normal and mutant avian integrin subunits in rodent cells. J Cell Biol 1989; 109: 853-861.
60. Hayashi Y, Haimovich B, Reszka A et al. Expression and function of chicken integrin β_1 subunit and its cytoplasmic domain mutants in mouse NIH 3T3 cells. J Cell Biol 1990; 110: 175-184.
61. Horwitz A, Duggan K, Buck C et al. Interaction of plasma membrane fibronectin receptor with talin - a transmembrane linkage. Nature 1986; 320: 531-533.
62. Tapley P, Horwitz A, Buck C et al. Integrins isolated from Rous sarcoma virus-transformed chicken embryo fibroblasts. Oncogene 1989; 4: 325-333.
63. Otey C, Pavalko FM, Burridge, K. An interaction between α-actinin and the β_1 integrin subunit in vitro. J Cell Biol 1990; 111: 721-729.
64. Haimovich B, Aneskievich BJ, Boettiger D. Cellular partitioning of β_1 integrins and their phosphorylated forms is altered after transformation by Rous sarcoma virus or treatment with cytochalasin D. Cell Regulation 1991; 2: 271-283.
65. Plantefaber LC, Hynes RO. Changes in integrin receptors in oncogenically transformed cells. Cell 1989; 56: 281-290.
66. Turner CE, Glenney JR, Burridge K. Paxillin: a new vinculin-binding protein present in focal adhesions. J Cell Biol 1990; 111: 1059-1068.
67. Turner CE. Paxillin: a cytoskeletal target for tyrosine kinases. BioEssays 1994; 16: 47-52.
68. Turner CE. Paxillin is a major phosphotyrosine-containing protein during embryonic development. J Cell Biol 1991; 115: 201-207.
69. Glenney JR. Tyrosine-phosphorylated proteins: mediators of signal transduction from tyrosine kinases.Biochim Biophys Acta 1992; 1134: 113-127.
70. Weng Z, Taylor JA, Turner CE et al. Detection of src-SH3 binding proteins, includ-

ing paxillin, in normal and *v-src*-transformed BALB/c 3T3 cells. J Biol Chem 1993; 268: 14656-14963.

71. Burridge K, Turner CE, Romer LH. Tyrosine phosphorylation of paxillin and pp125^{FAK} accompanies cell adhesion to extracellular matrix: a role in cytoskeletal assembly. J Cell Biol 1992; 119: 893-903.
72. Davis S, Lu ML, Lo SH et al. Presence of an SH2 domain in the actin-binding protein tensin. Science 1991; 252: 712-715.
73. Hanks SK, Calab MB, Harper MC, Patel SK. Focal adhesion kinase phosphorylated in response to cell attachment to fibronectin. Proc Natl Acad Sci 1992; 89: 8487-8491.
74. Schaller, MD, Borgman CA, Parsons JT. Autonomous expression of the non-catalytic domain of the focal adhesion associated protein tyrosine kinase pp125^{FAK}. Mol Cell Biol 1993; 13: 785-791.
75. Kornberg, LJ, Earp HS, Turner CE et al. Signal transduction by integrins: increased protein tyrosine phosphorylation caused by clustering of B1 integrins. Proc Natl Acad Sci 1991; 88: 8392-8396.
76. Guan JL, Trevithick JE, Hynes RO. Fibronectin/integrin interaction induces tyrosine phosphorylation of a 120-kDa protein. Cell Reg 1992; 2: 951-964.
77. Schaller MD, Borgman CA, Cobb BS et al. pp125^{FAK}, a structurally unique protein tyrosine kinase associated with focal adhesions. Proc Natl Acad Sci USA 1992; 89: 5192-5196.
78. Guan J-L, Shalloway D. Regulation of focal adhesion-associated protein tyrosine kinase by both cellular adhesion and oncogenic transformation. Nature 1992; 358: 690-692.
79. Lipfert L, Haimovich B, Schaller MD et al. Integrin-dependent phosphorylation and activation of the protein tyrosine kinase pp125^{FAK} in platelets. J Cell Biol 1992; 119: 905-912.
80. Huang MM, Lipfert L, Cunningham M et al. Adhesive ligand binding to integrin aIIbβ3 stimulates tyrosine phosphorylation of novel protein substrates before phosphorylation of pp125^{FAK}. J Cell Biol 1993; 122: 473-483.
81. Turner, CE, Schaller MD, Parsons JT. Tyrosine phosphorylation of the focal adhesion kinase pp125^{FAK} during development: relation to paxillin. J Cell Sci 1993; 105: 637-645.
82. Kanner SB, Reynolds AB, Vines RR et al. Monoclonal antibodies to individual tyrosine phosphorylated protein substrates of oncogene-encoded tyrosine kinases. Proc Natl Acad Sci USA 1990; 87: 3328-3332.
83. Cobb BS, Schaller MD, Leu TH et al. Stable association of pp60^{src} and pp59^{fyn} with the focal adhesion-associated protein tyrosine kinase, pp125^{FAK}. Mol Cell Biol 1994; 14: 147-155.
84. Romer LH, Burridge K, Turner CE. Signaling between the extracellular matrix and the cytoskeleton: tyrosine phosphorylation and focal adhesion assembly. Cold Spring Harbor Symp Quant Biol 1993; 62: 193-201.
85. Rankin S, Rozengurt E. Platelet-derived growth factor modulation of focal adhesion kinase (p125^{FAK}) and paxillin tyrosine phosphorylation in Swiss 3T3 cells. J Biol Chem 1994; 269: 704-710.
86. Akiyama T, Kadowaki T, Nishida E et al. Substrate specificities of tyrosine-specific protein kinases toward cytoskeletal proteins in vitro. J Biol Chem 1986; 261: 14797-14803.
87. Crumpton MJ, Bedman JR. Protein terminology tangle . Nature (London) 1990; 345: 212.
88. Glenney JR. Phosphorylation of p36 in vitro with pp60^{src}. FEBS Letts 1985; 192: 79-82.
89. Nakamura KD, Weber MJ. Phosphorylation of a 36,000Mr cellular protein in cells infected with partial transformation mutants of Rous sarcoma virus. Mol Cell Biol 1982; 2: 147-153.
90. Nermut MV, Eason P, Hirst EMA, Kellie S. Cell/substratum contacts in RSV-transformed cells. Exptl Cell Res 1991; 193: 382-387.
91. Wu H, Reynolds, AB, Kanner SB et al. Identification and characterization of a novel cytoskeleton-associated pp60^{src} substrate. Mol Cell Biol 1991; 11: 5113-5124.
92. Wu H, Parsons JT. Cortactin, an 80/85 kilodalton pp60^{v-src} substrate, is a filamentous actin-binding protein enriched in the cell cortex. J Cell Biol 1990; 120: 1417-1426.

93. Fukami Y, Nakamura T, Nakayama A, Karehisa T. Phosphorylation of tyrosine residues of calmodulin in Rous sarcoma virus transformed cells. Proc Natl Acad Sci USA 1986; 83: 4190-4193.
94. Reynolds AB, Herbert JL, Cleveland ST et al. P120, a novel substrate of protein tyrosine kinase receptors and of p60^{src}, is related to cadherin-binding factors β-catenin plakoglobin and armadillo. Oncogene 1992; 7: 2439-2445.
95. Kanner, SB, Reynolds AB, Wang H-CR et al. The SH2 and SH3 domains of pp60^{src} direct stable association with tyrosine-phosphorylated p130 and p110. EMBO J 1991; 10: 1689-1698.
96. Davis S, Lu ML, Lo S et al. Presence of an SH2 domain in the actin-binding protein tensin. Science 1991; 252: 712-715.
97. Rothberg KG, Henser JE, Donzell WC et al. Caveolin, a protein component of caveolae membrane coats. Cell 1992; 68: 673-682.
98. Schriver K, Rohrschneider L. Organization of pp60^{src} and selected cytoskeletal proteins within adhesion plaques and junctions of Rous sarcoma virus-transformed cells. J Cell Biol 1981; 89: 525-535.
99. David-Pfeuty T, Nouvian-Dooghe Y. Immunolocalization of the cellular *src* protein in interphase and mitotic NIH *c-src* overexpressor cells. J Cell Biol 1990; 111: 3097-3116
100. Marchisio PC, Cirilo D, Naldini L et al. Cell-substratum interaction of cultured avian osteoclasts is mediated by specific adhesion structures. J Cell Biol 1984; 99: 1696-1705.
101. Marchisi, PC, Cirillo D, Teti A et al. Rous sarcoma virus-transformed fibroblasts and cells of monocytic origin display a peculiar dot-like organization of cytoskeletal proteins involved in microfilament-membrane interactions. Exp Cell Res1987; 169: 202-214.
102. Gavazzi I, Nermut MV, Marchisio PC. Ultrastructure and gold immunolabelling of cell-substratum adhesions (podosomes) in RSV-transformed BHK cells. J Cell Sci 1989; 94: 85-99.
103. Akiyama SK, Nagata K, Yamada KM. Cell surface receptors for ectracellular matrix components. Biochim Biophys Acta 1987; 1031: 91-110.
104. Giancotti FG, Comoglio PM, Tarone G. A 135,000 molecular weight plasma membrane glycoprotein involved in fibronectin-mediated cell adhesion. Exp Cell Res 1986; 163: 47-62.
105. Chen W-T, Wang J, Hasegawa et al. Regulation of fibronectin receptor distribution by transformation, exogenous fibronectin and synthetic peptides. J Cell Biol 1986; 103: 1649-1661.
106. Horvath AR, Kellie S. Regulation of integrin mobility and cytoskeletal association in normal and RSV-transformed chick embryo fibroblasts. J Cell Sci 1990; 97: 307-315.
107. Mueller SC, KellyT, Dai MZ et al. Dynamic cytoskeleton-integrin associations induced by cell binding to immobilized fibronectin. J Cell Biol 1990; 109: 3455-3464.
108. Rodriguez-Fernandez JL, Geiger B, Salomon D et al. Suppression of tumorigenicity in transformed cells after transfection with vinculin cDNA. J Cell Biol 1992; 111: 427-438.
109. Rodriguez-Fernandez JL, Geiger B, Salomon D et al. Overexpression of vinculin suppresses cell motility in Balb/c 3T3 cells. Cell Motil Cytoskeleton 1992; 22: 127-134
110. Giancotti FG, Ruoslahti E. Elevated levels of α_5B_1 fibronectin receptor suppresses the transformed phenotype of chinese hamster ovary cells. Cell 1990; 69: 849-859.
111. Stossel TP, Chapponier C, Ezzell RM et al. Nonmuscle actin binding proteins. Annu Rev Cell Biol 1985; 1: 353-402.
112. Goldschmidt-Clermont PJ, Janmey PA. Profilin, a weak CAP for actin and RAS. Cell 1991; 66: 419-421.
113. Goldschmidt-Clermont PJ, Machesky LM, Doberstein SK, Pollard TD. The actin-binding protein profilin binds to PIP_2 and inhibits its hydrolysis by phospholipase C. Science 1990; 247:1575-1578.
114. Lassing I, Lindberg U. Evidence that the phosphatidylinositol cycle is linked to cell motility. Exp Cell Res 1988; 174: 1-15.
115. Stossel TP. From signal to pseudopod. J Biol Chem 1989; 264: 18261-18264.
116. Stossel TP. On the crawling of animal cells. Science 1993; 260: 1086-1094.
117. Adarem A. Signal transduction and the actin

cytoskeleton: the role of MARKS and profilin. Trends Bioch Sci 1992; 17: 438-443

118. Machesky LM, Pollard, TD. Profilin as a mediator of membrane-cytoskeleton communication. Trends Cell Biol 1993; 3: 381-385.

119. Waites GT, Graham IR, Jackson P et al. Mutually exclusive splicing of calcium-binding exons in chick alpha-actinin. J Biol Chem 1992; 267: 6253-6271.

120. Blackshear PJ. The MARCKS family of cellular protein kinase C substrates. J Biol Chem 1993; 268: 1501-1504.

121. Hartwig JH, Thelen M, Rosen A et al. MARCKS is an actin filament crosslinking protein regulated by protein kinase C and calcium-calmodulin. Nature (London) 1992; 356:618-622.

122. Joseph CK, Qureshi SA, Wallace DJ, Foster DA. MARCKS protein is transcriptionally down-regulated in v-src-transformed BALB/c 3T3 cells. J Biol Chem 1992; 267: 1327-1330.

123. Cotton PC, Brugge JS. Neural tissues express high levels of the cellular *src* gene product pp60$^{c\text{-}src}$. Mol Cell Biol 1983; 3: 1157-1162.

124. Brugge JS, Cotton PC, Queral AE,et al. Neurones express high levels of a structurally modified, activated form of pp60$^{c\text{-}src}$. Nature (London) 1985; 316: 554-557.

125. Golden A, Nemeth S, Brugge JS. Blood platelets express high levels on the pp60$^{c\text{-}src}$ tyrosine kinase activity. Proc Natl Acad Sci USA 1986; 83: 852-856.

126. Maness PF, Aubrey M, Shores CG et al. *c-src* gene product in developing rat brain is enriched in nerve growth cone endings. Proc Natl Acad Sci USA 1988; 85: 5001-5006.

127. Cheng N, Sayhoun N. The growth cone cytoskeleton. Glycoprotein association, calmodulin binding, and tyrosine/serine phosphorylation of tubulin. J Biol Chem 1988; 8: 3935-3942.

128. Pang DT, Wang JKT, Valtorta F et al. Protein tyrosine phosphorylation in synaptic vesicles. Proc Natl Acad Sci USA 1988; 85: 762-766.

129. Barnekow A, Jahn R, Schartl M. Synaptophasin: a substrate for the protein tyrosine kinase pp60$^{c\text{-}src}$ in intact synaptic vesicles. Oncogene 1990; 5: 1019-1024.

130. Matten WT, Aubry M, West et al. Tubulin is phosphorylated at tyrosine by pp60$^{c\text{-}src}$ in nerve growth cones. J Cell Biol 1990; 111: 1959-1970.

131. David-Pfeuty T, Nouvian-Dooghe Y. Immunolocalization of the cellular *src* protein in interphase and mitotic NIH *c-src* overexpressor cells. J Cell Biol 1990; 111: 3097-3116.

132. Kaplan JM, Varmus HE, Bishop JM. The *src* protein contains multiple domains for specific attachment to membranes. Mol Cell Biol 1990; 10: 1000-1009.

133. Redmond T, Brott BK, Jove R et al. Localization of the viral and cellular src kinases to perinuclear vesicles in fibroblasts. Cell Growth Differ 1992; 3: 567-576.

134. Ferrel JE, Noble JA, Martin GS et al. Intracellular localization of pp60$^{c\text{-}src}$ in human platelets. Oncogene 1990; 5: 1033-1036.

135. Anand R, Wilkinson JM, Kellie S. Localization of pp60$^{c\text{-}src}$ to the surface membrane of human platelets. Oncogene 1993; 8: 3013-3020.

136. Elmore MA, Anand R, Horvath AR, Kellie S. Tyrosine-specific posphorylation of gpIIIa in platelet membranes. FEBS Letts. 1990; 269: 283-287.

137. Horvath AR, Muszbek L, Kellie S. Translocation of pp60$^{c\text{-}src}$ to the cytoskeleton during platelet aggregation. EMBO J 1992; 11: 855-861.

138. Clark EA, Brugge J. Distribution of activated pp60$^{c\text{-}src}$ to integrin-dependent cytoskeletal complexes in thrombin-stimulated platelets. Mol Cell Biol 1993; 13: 1863-1871.

139. Findik D, Reuter C, Presek P. Platelet membrane glycoproteins IIb and IIIa are substrates of purified pp60$^{c\text{-}src}$ protein tyrosine kinase. FEBS Letts 1990; 262: 1-4.

140. Ferrell JE, Martin GS. Tyrosine-specific protein phosphorylation is regulated by by glycoprotein IIb-IIIa in platelets. Proc Natl Acad Sci USA 1989; 86: 2234-2238.

CHAPTER 5

The Interaction of Tyrosine Kinases with Other Signaling Molecules

The tyrosine-specific phosphorylation of several cytoskeletal proteins by $pp60^{v\text{-}src}$ was described in chapter 4. In order to phosphorylate these proteins the *src* gene product must interact with the cytoskeleton, however the specific cytoskeletal components to which $pp60^{v\text{-}src}$ binds have not yet been identified. Nevertheless over the last few years it has been recognized that $pp60^{v\text{-}src}$ and other tyrosine kinases associate with other signaling molecules within the cell. The first hint that oncogenic tyrosine kinases associated with other molecules came from observations of additional enzyme activities other than tyrosine phosphorylation associated with several tyrosine kinases.

ASSOCIATION OF $PP60^{V\text{-}SRC}$ WITH PHOSPHATIDYLINOSITOL-3'-KINASE

In addition to increased tyrosine phosphorylation, transformed cells commonly display changes in other signaling molecules, and one which has received much attention is the change in inositol phospholipid metabolism. Oncogenic transformation commonly leads to elevated levels of phosphatidylinositol (PI), phosphatidylinositol phosphates (PIP) and diacylglycerol (DAG), and evidence of enhanced PI kinase activity has been presented.[1-5] Early observations were that in addition to tyrosine kinase activity $pp60^{v\text{-}src}$ had the ability to phosphorylate these lipids in vitro.[6] When $pp60^{v\text{-}src}$ was immune precipitated from transformed cells the immunopurified protein phosphorylated PI, PIP, DAG and glycerol. Similar lipid kinase activities were found for the transforming protein of UR2, $p68^{v\text{-}ros}$ and so it was suggested that the overproduction of inositol phosphates and DAG in transformed cells was due to a direct effect of the transforming tyrosine kinase having dual specificity for lipid and protein.[7] Later investigations used more stringent biochemical purification of $pp60^{v\text{-}src}$ to show that the tyrosine kinase activity and the lipid kinase activity could be separated.[8] Others showed that removal of the tyrosine kinase activity from RSV, UR2 Fujinami virus or Abelson murine leukemia virus-transformed cells had no effect on the PI kinase activity,[9,10] thus the lipid kinases and protein kinases were different entities. It is now well accepted that these initial results were due to the association of phosphatidylinositol-3'-kinase with pp60$^{v\text{-}src}$ and the molecular mechanisms involved in this interaction will be described later.[11,12]

THE ASSOCIATION OF PP60$^{C\text{-}SRC}$ WITH POLYOMA VIRUS MIDDLE T ANTIGEN

Polyoma virus transforms rodent cells by the production of a specific viral protein, middle T antigen.[13,14] Initial studies into the mechanism of transformation revealed that middle T was a plasma membrane-associated protein which had an intrinsic tyrosine kinase activity.[15] Courtneidge and Smith[16,17] investigated this activity and found that the tyrosine kinase activity was similar to that of pp60$^{c\text{-}src}$. Further examination revealed that pp60$^{c\text{-}src}$ and middle T formed a stable complex and

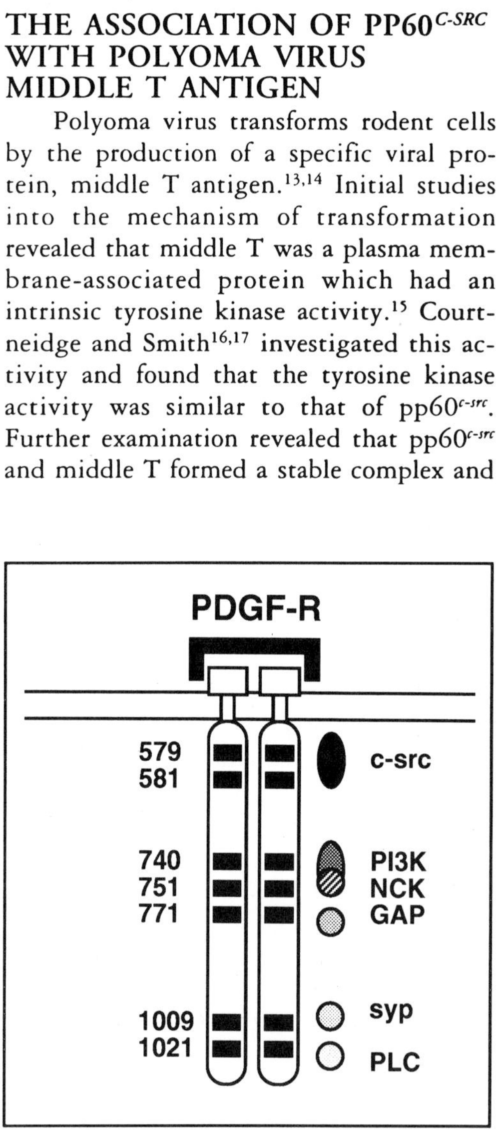

Fig. 5.1. The association of SH2-containing proteins with different tyrosine phosphorylated sites of the activated human platelet derived growth factor receptor.

it was established that the tyrosine kinase activity was due to pp60$^{c\text{-}src}$. However polyoma virus-transformed cells exhibited higher pp60$^{c\text{-}src}$ kinase activity than non-transformed cells, although the amount of pp60$^{c\text{-}src}$ protein was similar, suggesting that there was a change in the specific activity of pp60$^{c\text{-}src}$ after transformation.[18] In a careful study of this kinase activity Courtneidge[19] found that when pp60$^{c\text{-}src}$ was complexed to middle T the V_{max} of the enzyme kinase activity was about an order of magnitude higher. This elevated activity was shown to be due to the dephosphorylation of tyrosine 527 in pp60$^{c\text{-}src}$[20] (Fig. 5.2). The role of this residue in the regulation of pp60$^{c\text{-}src}$ kinase activity has now been extensively examined, and a kinase which specifically phosphorylates this residue, *c-src* kinase, has been cloned.[21,22] Investigations into the function of *c-src* kinase have shown that in addition to regulating the activity of *c-src*, it can also regulate the kinase activity of other members of the *src* family, and in particular is important in the pp56^{lck}-mediated regulation of T cell responses.[23,24] The ability of *c-src* kinase to suppress *c-src* activity requires both the SH2 and SH3 domains of *src*, possibly due to masking of tyr-527 by SH2/SH3-dependent protein folding.[25,26]

By analogy with pp60$^{v\text{-}src}$, the middle T-pp60$^{c\text{-}src}$ complex was found to have phosphatidylinositol kinase activity,[27] and anti-middle T or anti-pp60$^{c\text{-}src}$ immunoprecipitations from polyoma virus-transformed cells had 20- to 50-fold more PI kinase activity those from nontransformed cells. However in agreement with the findings for pp60$^{c\text{-}src}$ highly purified preparations of the middle T-pp60$^{c\text{-}src}$ complex had very little PI kinase activity, thus the activity originally detected was due to a PI kinase associating with the complex.[28] At least two PI kinase activities have now been identified in fibroblasts:[29] Type I is the major activity and was shown to have a novel enzyme activity which catalyzes phosphorylation of phosphatidylinositol in the 3' position of the inositol ring.[30] This is a ubiquitous enzyme which can also phos-

and $PI3,4,5P_3$. Although the function of these products is unknown it seems likely that the major product of PI3kinase activation within cells is PIP_3, which is not hydrolyzed by phospholipase C and therefore is a distinct putative second messenger from $PI4,5P_2$.[31] Otsu et al[32] purified PI3kinase from bovine brain and found that it consisted of two subunits of 85 kDa and 110 kDa. These subunits were cloned and when expressed in fibroblasts the p85 subunits had the ability to bind to the PDGF and EGF receptors after stimulation. p85 could also associate with middle T-*c-src* complexes, and the intermolecular associations could be mimicked in vitro by expressing the proteins in a baculovirus system which allowed the production of native protein. The 110 subunit is the catalytic component of the enzyme, and associates with p85 via a sequence between the two SH2 domains of p85.[33,34] The structural requirements for pp60-PI kinase association will be described later.

In addition to changes in enzyme activity when $pp60^{c\text{-}src}$ is complexed to middle T there is also a change in subcellular distribution of the enzyme. Whilst $pp60^{v\text{-}src}$ associates with the detergent-insoluble cytoskeleton, very little $pp60^{c\text{-}src}$ associates with the cytoskeleton and little is localized to adhesion plaques.[35,36,37] The molecular mechanism for this difference is unknown, although the activation of the cellular protein by dephosphorylation of tyrosine 527 is almost certainly involved. After polyoma virus transformation, however, $pp60^{c\text{-}src}$/middle T complexes associate with the detergent-insoluble cytoskeleton and localize in focal contacts in a manner similar to that found for $pp60^{v\text{-}src}$.[38] Furthermore a physical association with this complex and the adhesion plaque protein vinculin was detected. These results are all consistent with one mode of action of polyoma virus transformation by middle T-induced activation of $pp60^{c\text{-}src}$, resulting in an enhanced tyrosine kinase activity and a different subcellular localization similar to that of the transforming kinase $pp60^{v\text{-}src}$.

It is now clear that oncogenic tyrosine kinases such as *src* associate with a number of different intracellular molecules, and the past 2-3 years has seen an explosion in this field with an ever increasing number of tyrosine kinase-binding proteins being reported. The interactions of *src* with signaling molecules within the cell are similar to the interactions of tyrosine kinase growth factor receptors, and these signaling pathways overlap heavily, therefore I will discuss both of these classes of molecules together wherever possible in this chapter. The regions of tyrosine kinases which interact with other proteins have now been well defined in terms of sequence and domain structure.

SH2 DOMAINS

The family of cytoplasmic, *src*-related tyrosine kinases is still expanding, and sequence comparisons between these and also between other signaling molecules led to the discovery of sequence homologies which are conserved in many signaling proteins. The first of these homologies to be properly defined was the *src* homology 2 (SH2) domain, based on similarities with a non-catalytic N-terminal region of *src*. SH2 domains consist of sequences of about 100 amino acids. Subsequently another homologous sequence was defined and termed the SH3 domain and this will be described later. More than 20 proteins have now been shown to have SH2 domains, and most of these are postulated to play some role in intracellular signaling (Fig. 5.1). Many signaling proteins also contain SH3 domains and it appears that these proteins can be categorized either as catalytic proteins (e.g. kinases), structural molecules (e.g. cytoskeletal proteins), adaptor molecules linking SH2 and SH3-containing proteins together, or response molecules involved in cellular responses to stimuli as depicted in Table 5.1.

The role of SH2 domains in regulating protein-protein interactions during activation of the PDGF or EGF receptors have been studied most extensively, however similar associations occur when cell are stimulated by a number of other growth factors including colony stimulating factor 1, insulin or transformation by oncogenic tyrosine kinases. Mutations in the SH2 domains of *src*, *fps*, and *abl* modulate the specificity and activity of these proteins indicating that these domains are regulatory regions.[39]

Table 5.1. Examples of proteins containing SH2 and SH3 domains

	SH2 or SH3 Domain
Catalytic Molecules	
src	Both
PI3 kinase	Both
Phospholipase Cγ	Both
GTPase Activating factor	Both
syk	SH2
Phosphotyrosine phosphatase 1C	SH2
Myosin	SH3
Structural Molecules	
Tensin	SH2
Spectrin	SH3
Cortactin	SH3
Adaptor Molecules	
crk	Both
grb2	Both
shc	SH2
nck	Both
Response Molecules	
$p47^{phox}$	SH3
$p67^{phox}$	SH3
$p40^{phox}$	SH3

Table 5.2. Examples of SH2 binding motifs

SH2 Domain	Sequence
p85a	YMDM, YVPM
src	YIYV, YEEI
GAP	YNAP
ABL	YENP
PLCγC	YIIP
PLCγN	YLDL
GRB2	YVNI, YINQ, YLNT
SHC	YLRV

SH2 Domain Binding Specificity

The SH2 domains of a number of proteins have been cloned and expressed in several cell expression systems. However by far the most common method for generating SH2 domains to investigate binding interactions has been to express the SH2 domain as a fusion protein containing glutathione-S-transferase (GST) in *E. coli*. This protein can bind with high affinity to glutathione covalently linked to agarose beads to provide an SH2 domain affinity matrix.[40]

SH2 domains from a variety of proteins have been shown to bind to several tyrosine phosphorylated proteins including PLCγ, *ras*GAP and *abl*.[41-45] Comparison of the sequences of PDGF receptor or polyoma middle T protein which bind to the SH2 domain of PI3'kinase showed that the sequences surrounding the phosphotyrosine residue are important in both the binding and the specificity of the interaction, for example the sequence pYVPML has been found to inhibit the interaction of PI3'kinase with the PDGF receptor but does not inhibit the interaction of *ras*GAP or PLCγ.[46-51] Songyang et al[52] extensively investigated the binding specificity of tyrosine phosphorylated peptides to several SH2 domains using a degenerate peptide library in which any of the three residues downstream of the phosphotyrosine were randomly specified. This showed that the PI3'kinase SH2 domain had specificity for methionine three residues downstream of the phosphotyrosine whereas the carboxy-terminal PLCγ SH2 had selectivity for pYLIP or pYIIP and the *src* SH2 selected pYEEI. Other SH2 domains demonstrated some variability in their binding selectivities, for example Grb2 showed preference for asparagine in the +2 position after the phosphotyrosine but less selectivity for the +1 and +3 positions (See Table 5.2 for other examples).

Kinetic Parameters of SH2-Phosphopeptide Association

SH2 domain binding to phosphopeptide sequences can occur with relatively high affinities. Using a comparatively new technique which gives real-time kinetics and affinity measurements detected by a biosensor apparatus (BIACORE), several groups have quantitated the kinetic parameters of SH2 domain-phosphopeptide binding. The interaction of the p85 subunit of PI3'kinase with a phosphopeptide containing the sequence YMXM (derived from a major insulin receptor substrate IRS-1) was found to be high affinity with dissociation constant of 0.3-3 nM and very high on rates and off rates.[53] Similar affinity constants, high association rates and fast dissociation rates were found for p85α, GAP and PLCγ SH2 domain-phosphopeptide binding although some specificity in the reactions was also detected.[54] These workers suggested that the rapid on/off interactions would allow other phosphorylation sites within the SH2 domain-containing molecules themselves to interact with the SH2 domains in both intra- and inter-molecular ways, resulting in interaction with other downstream molecules, deactivation or other regulatory events.

SH2 Domain Structure

The structure of a number of SH2 domains have now been solved by both high resolution NMR and X-ray crystallography.[55-57] The SH2 domain of pp60*src* consists of two antiparallel β sheets surrounded by a helices. When an 11 residue phosphotyrosine-containing peptide containing YEEI binds and makes contact with the SH2 domain the four residues: phosphotyrosine, glu(+1), glu(+2) and ile(+3), interact such that the peptide fits snugly on to the surface of the SH2 domain. There is also a hydrophobic binding pocket surrounding the ile(+3), and this is conserved in many SH2 domains.[58-60] Binding of the phosphopeptide induces a small conformational change on the SH2 domain of *src*, but a much larger conformational change in the SH2 domain of p85.[61-63]

Molecular Analysis of Interactions between Signaling Molecules

Using a combination of transfection of PDGF receptor mutants, binding analysis and cellular responses, a number of SH2 domain binding sites have been identified. The SH2 binding regions of the PDGF receptor have been localized to regions along the length of the cytoplasmic domain (Fig. 5.1). The region close to the plasma membrane contains the autophosphorylation site when the receptor is occupied and deletions or site-directed mutagenesis have shown that Y579 and Y581 contain the binding sites for *src* proteins.[64-66] A region comprising an insert in the kinase domain contains binding sites for PI3' kinase, *nck* and *ras*GAP. Y740 and Y751 both bind to PI3'kinase such that each associates with a different SH2 domain of the p85α subunit. Y751 also binds to the SH2 domain of *nck*, a protein which has both SH2 and SH3 domains.[67] When the PI3' kinase binding sites were mutated or deleted there was a lack of PDGF-stimulated proliferation and PI3'kinase was not activated although other responses such as calcium mobilization were unaffected, showing that the association of PI3'kinase with the PDGF receptor is essential for proliferation.[68] In the carboxy tail of the PDGF receptor Y1009 and Y1021 bind to PLCγ and *syp*.[69,70,71] When Y1021 was replaced with phenylalanine PLCγ bound only weakly to the PDGF receptor and hydrolysis of PIP_2 was not stimulated. These experiments demonstrate that these protein interactions have specificity are functionally important in cell signaling. A more elaborate series of experiments was performed by Vallius and Kaslauskas[72] who mutated individually the binding sites for PI3 kinase, GAP, PLCγ and a 64 kDa protein in the PDGF receptor. These experiments indicated that the proteins which associate with activated PDGF receptor and which are required for initiation of DNA synthesis are either PI3kinase or PLCγ, but not GAP or the 64 kDa protein.

Similar mutation experiments have been performed with the fibroblast growth factor (FGF) receptor. A short sequence of amino acids surrounding tyrosine 766 is a binding site for the SH2 domain of PLCγ, however if this tyrosine is mutated to phenylalanine then PLCγ will not associate with the FGF receptor. This mutant receptor also cannot mediate phosphatidylinositol hydrolysis or calcium mobilization but will still mediate mitogenesis.[73,74] Phosphatidylinositol-3'-kinase appears to play a central role in many cell responses to stimulating agents.[75] In addition to binding to PDGF receptor, it has also been shown that GM-CSF stimulation of myeloid cells results in the activation of PI3'kinase and the formation of tyrosine kinase-containing signaling complexes which include PI3'kinase.[76] PI3'kinase is also activated and forms a complex with pp60$^{c\text{-}src}$ in the cytoskeleton of activated platelets.[77-79] The central sole of PI3'kinase in signaling responses has been further highlighted by the discovery that the fungal metabolite wortmannin, originally thought to be an inhibitor of myosin light chain kinase (at micromolar concentrations), is a potent and specific inhibitor of PI3'kinase at nanomolar concentrations. This agent appears to inactivate PI3'kinase by irreversibly binding to the 110 kDa subunit of PI3'kinase. There have only been a limited number of studies using wortmannin to inhibit PI3'kinase, however in these cases inactivation of PI3'kinase by nanomolar concentrations of wortmannin has led to inhibition of Fc receptor-mediated histamine release from basophils,[80] changes in fMLP-induced actin in neutrophils,[81] and inhibition of PDGF-mediated actin rearrangements in fibroblasts.[82] Although there have not yet been any reports on the use of wortmannin in transformed cells, this could be a powerful tool in allowing the analysis of signaling molecules necessary for the maintenance of transformation by oncogenic tyrosine kinases.

SH2 and *src* Regulation

The *src* family now consists of about 14 members which, in addition to the catalytic, SH2 and SH3 domains, all have conserved sequences at the carboxy terminus. As previously described the kinase activity of pp60$^{c\text{-}src}$ can be activated by a number of events including middle T binding or dephosphorylation. Mutations which activate pp60$^{c\text{-}src}$ have been variously mapped to the kinase domain, the SH2 and SH3 domains, and the carboxy terminus, suggesting that these regions all interact with each other to regulate kinase activity.[83] Courtneidge originally showed that treatment of pp60$^{c\text{-}src}$ with phosphatases activated the enzyme activity[19] and subsequently it was established that there is an essential tyrosine residue (tyr527) which regulates activity.[84-90] If tyr527 is phosphorylated pp60$^{c\text{-}src}$ kinase activity is low whereas if tyr527 is dephosphorylated the enzyme activity is high.[91] A model has been presented in which the inactive molecule is folded such that the carboxy terminus containing tyr527 interacts with the SH2 domain, and this has recently been modified to include the interaction with parts of the SH3 domain.[25] The result of this folding is to mask the catalytic domain of pp60$^{c\text{-}src}$ and thus maintain enzyme repression. When activated by dephosphorylation, tyr527 no longer interacts with the SH2 domain, since these domains bind tightly only to phosphotyrosine-containing peptides, exposing the catalytic domain and therefore allowing access of substrates to the enzyme site (Fig. 5.2).[92,93] C-terminal *src* kinase (CSK), the enzyme which phosphorylates tyr527 has been cloned and found to phosphorylate all members of the *src* family in a similar manner, and it seems likely that within cells CSK contributes to the maintenance of low activity pp60$^{c\text{-}src}$.[22-24] It is assumed that there is a phosphatase which dephosphorylates tyr527 to cause activation of pp60$^{c\text{-}src}$, however this has not yet been identified. CSK and the phosphatase may be important in cells where there is a deregulation of *c-src* activity, such as colon carcinoma as will be discussed in chapter 6. When CSK is overexpressed it inhibits the transformation of fibroblasts by *src*[94] and it may also play an important role in normal development since transgenic mice with CSK deletions have neural tube defects and circulatory lesions with high *src*-associated tyrosine kinase activities, which may be related to the high endogenous levels of pp60$^{c\text{-}src}$ in neurons and blood platelets.[95,96]

SH3 DOMAINS

SH3 domains are regions of amino acid homology of about 50-60 residues. Unlike SH2 domains SH3 domains do not bind to phosphotyrosine-containing sequences but to regions containing proline residues as will be described later. Many proteins contain SH3 domains in addition to SH2 domains, and their juxtaposition has suggested that SH3 domains are also involved in protein-protein interactions, however there is also a subset of proteins which contain only SH3 domains. Early studies indicated that SH3 domains might play a role in the association of proteins with cytoskeletal elements, since many of the SH3-containing proteins were cytoskeletal or associated with the cytoskeleton depending on their activation state. For example spectrin, myosin 1, and two components of the phagocyte NADPH oxidase all have SH3 domains and all associate with the cytoskeleton.[97] Furthermore mutations in the SH3 domain of pp60$^{v\text{-}src}$ led to the inability of this protein to interact with the cytoskeleton and a reduction in its transforming ability.[98] Evidence that SH3 domains are involved in cytoskeletal association has been presented by Bar-Sagi,[99] who found that when the isolated SH3 domain of PLCγ was microinjected into fibroblasts, this domain localized to the actin cytoskeleton. Other SH3-containing proteins tend to be signaling molecules such as PLCγ, p85α, and rasGAP. Another set of proteins exemplified by Crk, Grb2 and Nck are involved in intracellular signaling but contain no enzymatic sequences.[100]

Binding Specificities of SH3 Domains

Compared to SH2 domains, less is known about the structures to which SH3 domains bind, and the relevance of this to intermolecular interactions. In the last two years, however, a consensus of amino acid sequences which bind to SH3 domains has begun to emerge. The first protein to be identified as an SH3 binding protein was a protein designated 3BP1. This was isolated by screening a human placenta λgt11 expression library with the isolated SH3 domain of *abl* expressed as a GST fusion protein and selecting clones which bound the SH3 domain.[101] The amino acid sequences to which abl SH3 bound were then

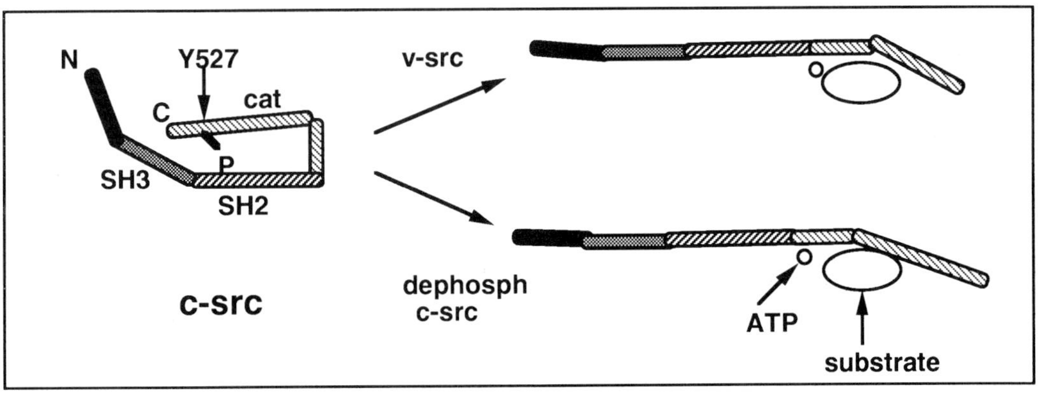

Fig. 5.2. Activation of the tyrosine kinase activity in pp60$^{v\text{-}src}$ and pp60$^{c\text{-}src}$. pp60$^{c\text{-}src}$ normally has tyrosine 527 phosphorylated and the protein folds over such that this phosphotyrosine residue associates with the SH2 domain. When tyrosine 527 is dephosphorylated, or in pp60$^{v\text{-}src}$ which lacks this residue, the protein changes conformation such that tyrosine 527 no longer associates with the SH2 domain, unmasking the catalytic domain.

analyzed by truncating the cloned SH3-binding protein and a short sequence of amino-acids which specifically bound to the *abl* SH3 domain was isolated. The sequence APTMPPPLPP was found to be the minimal binding sequence.[102] The function of 3BP1 is still unknown although the protein contains some homology to rhoGAP and also has similarity to a region of bcr, a protein involved in the etiology of chronic myeloid leukemia (chapter 6). Subsequently it was found that an additional protein essential for cell signaling, Sos, bound to the SH3 domain of Grb2, PLCγ, *src*, *fgr*, *fyn* and *crk* via a proline-rich amino acid sequence.[103] Yet another SH3-binding protein, dynamin, has been identified.[104] Dynamin binds only a subset of SH3 domains such those from p85α, PLCγ, and *src*, but not to others such as those from p47*phox* or p67*phox*. The sequence on dynamin which bound to the SH3 domains was a long sequence containing two potential binding motifs of APAVPPARPGS or GAPPVPSRPGA. Interestingly, the binding of dynamin to an SH3 domain stimulated the GTPase activity of dynamin and this may have functional consequences since dynamin has been postulated to play a role in intracellular vesicle transport.[104-106] Several other SH3 binding proteins have now been characterized, although in most cases

Table 5.3. SH3 domain binding motifs

Protein	Sequence
3BP-1	PTMPPPLPPVP
3BP-2	PAYPPPPVPVP
DYNAMIN	APAVPPARPGS
DYNAMIN	GAPPVPSRPGA
SOS	EVPVPPPVPPR
SOS	HLDSPPAIPPR
SOS	HSIAGPPVPPR
CDC42	APKPMPPRPPL
P85	PPRPLPVAPGS
P85	PAPALPPKPPK
p67phox C	PQPAVPPRPSA

This table is not meant to be extensive survey of all SH3-binding domains, but should illustrate the conserved features.

the binding sequences have yet to be identified: paxillin and P85α bind to the SH3 domain of *src*,[107,108] and the GTPase activating protein CDC42 binds to P85α and *src* SH3 domains.[109] Recently Yu et al[110] have performed an extensive survey of p85α SH3-binding peptides and suggested a consensus sequence of XpΦPpXP where X represents nonconserved amino acids, p represents amino acids that tend to be prolines and I represents hydrophobic residues. In leucocytes the NADPH oxidase component p47*phox* binds to the carboxy-terminal SH3 domain of p67*phox* via a proline-rich sequence QPAVPPRP (P. Finan in press). Whilst there is an ever-increasing number of SH3-binding proteins being identified, the regulation of binding is still unclear. With SH2 domains the binding can be regulated by phosphorylation/dephosphorylation of tyrosine residues, however such a direct regulation does not occur in the interaction of SH3 domains with proline-rich sequences. It is possible that other post-translational modifications of SH3 domain-containing proteins regulate binding to proline-rich peptides. For example, p47*phox* is normally a cytosolic protein which translocates to the plasma membrane in an SH3 domain-dependent manner when it is phosphorylated by protein kinase C.[111-113] The phosphorylation site of this protein is without the SH3 domains, so it is possible that serine/threonine phosphorylation unmasks the SH3 domain to allow p47*phox* to bind to its specific proline-rich sequence. Alternatively phosphorylation could unmask the SH3 binding motif rather than the SH3 domain itself. Recently it has been found that the SH3 domain of the *lyn* and *fyn* tyrosine kinases bound to a proline-rich sequence in the 85 kDa subunit of PI3 kinase. This binding resulted in a fivefold increase in PI3 kinase activity and so one mechanism for PI3 kinase activation is via SH3 domain binding.[114] Some examples of the binding motifs for several SH3 domains are listed in Table 5.3. Further questions regarding the functional relationships between enzyme activation and SH3 domain binding have yet to be answered.

SH3 Domain Structure

Like SH2 domains, the structure of several SH3 domains have been solved using both NMR and X-ray crystallography[115-118] and the overall structures are similar. The *src* SH3 domain structure consists of a pair of three stranded anti-parallel β sheets which form a barrel-like structure. A hydrophobic pocket is formed by a number of conserved aliphatic and aromatic residues into which fits the proline-rich binding motif. Yu et al[119] have obtained evidence for two proline-binding sites in the SH3 domain of p85α: a primary site formed by the conserved residues Tyr14, Trp55, Pro70 and Tyr73 and a secondary site formed partly by Tyr12 on which the proline peptide adopts an extended conformation in which all the prolines are in *trans* configuration with the peptide forming a left-handed type II polyproline helix. NMR spectroscopy has shown that the SH3 domain of GAP has a similar structure to that of other SH3 domains, and has two accessible hydrophobic domains which could function as binding sites, which may be the reason why SH3-containing proteins can bind to several proteins. The residues 317-326 of GAP which are essential for *ras* signaling are in the SH3 domain and form two lipophilic bulges near the hydrophilic domains and may therefore play a role in protein recognition.[120]

The Pathways of Signaling by SH2 and SH3 Domain-Containing Proteins

Much of the initial investigations of the signaling pathways which involve SH2 and SH3 domains used *Drosophila* mutants to map the functional consequences of deletions or mutations in specific genes. The activation of *ras* coupled to serine/threonine phosphorylation appears to play an essential role in growth factor-stimulated proliferation of mammalian cells and also possibly in transformation by tyrosine kinases. Several key players in this pathway have now emerged: Grb2, Sos, Shc and ras itself. Sos is a protein originally characterized as an essential component for *Drosophila* and *C. elegans* differentiation.[21-127] Biochemical analysis has revealed that Sos is a guanine nucleotide-releasing factor for the *ras* GTPase. Grb2 is a protein which contains only one SH2 and two SH3 domains, and this protein enhances Sos guanine nucleotide-releasing activity. When mammalian cells are stimulated by EGF there is autophosphorylation of the receptor, which then induces binding of Grb2 at Tyr1068. Grb2 can also bind to Sos via its SH3 domains, which recognize a proline-rich motif on Sos. Thus after EGF stimulation the net effect is that Sos is recruited to the plasma membrane in a complex with Grb2, which places it in close proximity to *ras* where it can influence the ras GTPase activity.[128-132] When cells are transformed by oncogenic tyrosine kinases such as *src*, another molecule also comes into play. *v-src* transformation leads to tyrosine phosphorylation of Shc, an SH2 domain containing protein[133,134,135] and indeed overexpression of Shc can result in the transformation of fibroblasts.[134,135] Tyrosine phosphorylation of Shc induces its association with the SH2 domain of Grb2, leading to the formation of a Shc/Grb2/Sos complex. This then couples to *ras* activation, again possibly by affecting *ras* GTPase activity. The functional importance of Shc-containing complexes is underscored by the formation of similar complexes when cells are stimulated with insulin or fibroblast growth factor.[136-138] In summary tyrosine phosphorylation either by growth factors or by oncogenic tyrosine kinases stimulate similar activation pathways involving intermolecular associations of SH2 domain and SH3 domain-containing proteins. In addition to coupling growth factor signaling to proliferation, Grb 2 also couples EGF and PDGF receptor occupancy to cytoskeletal reorganization. If an antibody to Grb2 is microinjected into fibroblasts the reorganization of the actin-containing cytoskeleton by PDGF or EGF was abolished, suggesting that in addition to influencing ras activity in cells Grb2 also plays a role in the pathways leading to rac

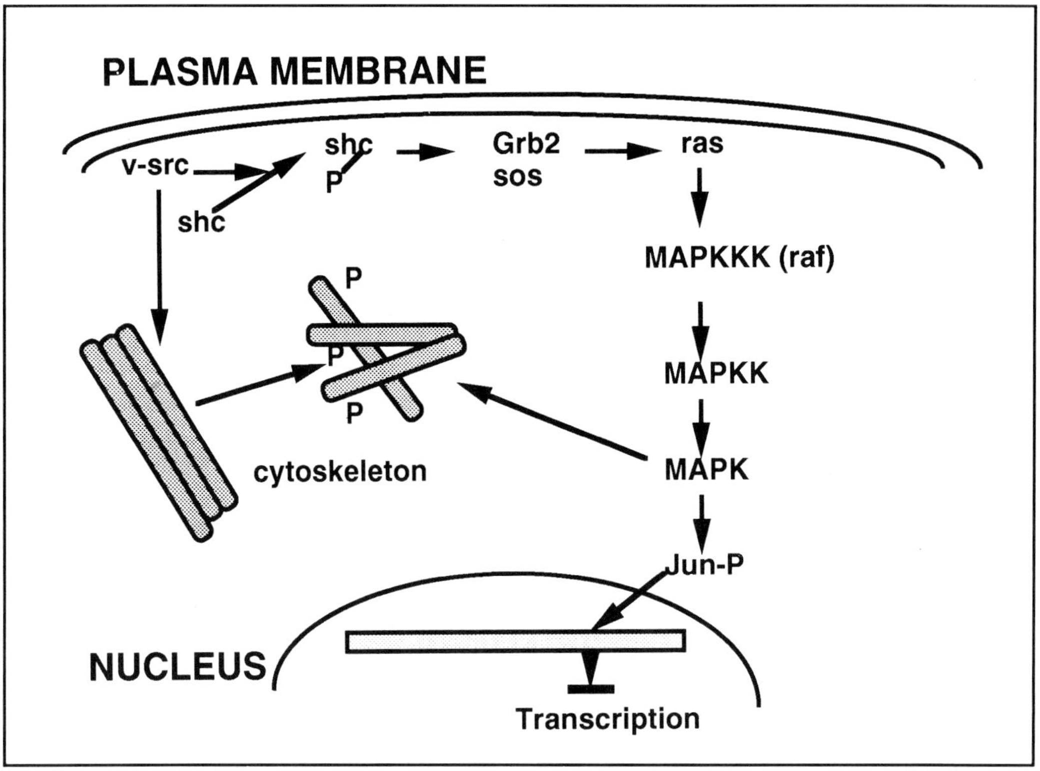

Fig. 5.3. Possible signaling pathways leading to neoplastic transformation by pp60^v-src. pp60^v-src phosphorylates SHC which then associates with Grb2/sos complex, activating ras *by enhancing nucleotide exchange. This then leads to expression of new genes by phosphorylation of transcription factors such as* jun *via the MAP kinase pathway. Concurrently, pp60^v-src may directly phosphorylate other substrates such as cytoskeletal proteins leading to the characteristic morphological and cytoskeletal changes, or phosphorylate rhoGAP leading to rho-mediated cytoskeletal rearrangements.*

and rho activity.[139] Grb2 associates with dynamin through its SH3 domain and it is likely that this plays a role in any cytoskeletal events related to vesicle fusion in cells.[140,141] A number of possible signaling pathways involved in the biochemical and ultrastructural changes associated with *src* transformation are depicted in Figure 5.3.

References

1. Auger KR, Cantley LC. Novel polyphosphoinositides in cell growth and activation. Cancer Cells 1991; 3: 263-270.
2. Bansal VS and Majerus PW. Phosphatidylinositol-derived precursors and signals. Ann Rev Cell Biol 1990; 6: 41-67.
3. Cantley LC, Auger KR, Carpenter C et al. Oncogenes and signal transduction. Cell 1991; 64: 281-302.
4. Berridge MJ, Irvine RF. Inositol phosphates and cell signaling. Nature (London) 1989; 341: 197-205.
5. Tones MA, Kellie S, Hawthorne JN. Elevated phosphatidylinositol kinase activity in Rous sarcoma virus-transformed cells. Lack of evidence for enzyme translocation. Biochim Biophys Acat 1987; 931: 165-169.
6. Sugimoto, YM, Whitman M, Cantley LC, Erikson RL. Evidence that the Rous sarcoma virus transforming gene product phosphorylates phosphatidylinositol and diacylglycerol. Proc Natl Acad Sci USA 1984; 81: 2117-2121.
7. Macara IG, Marinetti GV, Balduzzi PC. Transforming protein of avian sarcoma virus UR2 is associated with phosphatidylinositol kinase activity: possible role in tumorigenesis. Proc Natl Acad Sci USA

1984; 81: 2728-2732.

8. MacDonald, ML, Kuenzel EA, Glomset JA et al. Evidence from two transformed cell lines that the phosphorylation of peptides tyrosine and phosphatidylinositol are catalysed by different proteins. Proc Natl Acad Sci USA 1985; 82: 3993-3997.
9. Sugano S, Hanafusa H. Phosphatidylinositol kinase activity in virus-transformed and non-transformed cells. Mol Cell Biol 1985; 5: 3194-3198.
10. Fry MJ, Gebhradt A, Parker P, Foulkes JG. Phosphatidylinositol turnover and transformation of cells by Abelson murine leukemia virus. EMBO J 1985; 4: 3173-3178.
11. Fukui Y, Kornbluth S, Jong SM. Phosphatidylinositol kinase type I activity associates with various oncogene products. Oncogene 1989; 4: 283-292.
12. Fukui Y, Hanafusa H. Phosphatidylinositol kinase activity asociated with viral pp60^{src} protein. Mol Cell Biol 1989; 9: 1651-1658.
13. Rassoulzadegan M, Cowie A, Carr A et al. The roles of individual polyoma virus early proteins in oncogenic transformation. Nature 1982; 300: 713-718.
14. Treisman R, Novak U, Favaloro J et al. Transformation of rat cells by an altered polyoma virus genome expressing only the middle T protein. Nature 1981; 292: 595-600.
15. Schaffenhausen BS, Benjamin TL. Protein kinase activity associated with polyoma virus middle T antigen. Cold Spring Harbor Conf. Cell Proliferation 1981; 8: 1281-1298.
16. Courtneidge SA, Smith AE. Polyoma virus transforming protein associates with the product of the *c-src* cellular gene . Nature 1983; 303: 435-439.
17. Courtneidge SA, Smith AE. The complex of polyomavirus middle T antigen and pp60$^{c\text{-}src}$. EMBO J 1984; 3: 585-591.
18. Grussenmeyer T, Scheidtmann KH, Hutchinson MA et al. Complexes of ployoma virus medium T antigen and cellular proteins. Proc Natl Acad Sci USA 1985; 82: 7952-7954.
19. Courtneidge SA. Activation of the pp60$^{c\text{-}src}$ kinase activity by middle T antigen binding or by dephosphorylation. EMBO J 1985; 4: 1471-1477.
20. Yonemoto W, Jarvis-Morar M, Brugge JS et al. Tyrosine phosphorylation within the amino-terminal domain of pp60$^{c\text{-}src}$ molecules associated with polyoma virus middle-sized antigen. Proc Natl Acad Sci USA 1985; 82: 4568-4572.
21. Nada S, Okada A, MacCauley A et al. Cloning of a complementary DNA for a protein-tyrosine kinase that specifically phosphorylatres a negative regulatory site of pp60$^{c\text{-}src}$. Nature 1991; 35: 69-72.
22. Okada M, Nakagawa H. A protein tyrosine kinase involved in the regulation of pp60$^{c\text{-}src}$ function. J Biol Chem 1989; 264: 20886-20893.
23. Okada M, Nada S, Yamanashi Y et al. CSK: a protein-tyrosine kinase involved in the regulation of *src* family kinases. J Biol Chem 1991; 264: 24249-24252.
24. Bergman M, Mustelin T, Oetkin C et al. The human p50^{csk} tyrosine kinase phosphorylates p56^{lck} at tyr-505 and down-regulates its catalytic activity. EMBO J 1992; 11: 2919-2924.
25. Superti-Furga G, Furnagelli S, Koegl M et al. Csk inhibiton of *c-src* activity requires both the SH2 and SH3 domains of *src*. EMBO J 1993; 12: 2625-2634.
26. Murphy SM, Bergman M, Morgan DO. Suppression of *c-src* activity by c-terminal *src* kinase involves the *c-src* SH2 and SH3 domains: Analysis with *Saccharomyces cerevisiae.* Mol Cell Biol 1993; 13: 5290-5300.
27. Whitman M, Kaplan DR, Schaffhausen B et al. Association of phosphatidylinositol kinase activity with polyoma middle T competent for transformation. Nature 1985; 315: 239-242.
28. Koch W, Carbone A, Walter G. Purified polyoma virus medium T antigen has tyrosine-specific protein kinase activity but no significant phosphatidylinositol kinase activity. Mol Cell Biol1986; 6: 1866-1874.
29. Whitman M, Kaplan DR, Roberts TM, Cantley L. Evidence for two distinct phosphoinositide kinases in fibroblasts. Biochem J 1987; 247: 165-174.
30. Whitman M, Downes CP, Keeler M et al. Type I phosphatidylinositol kinase makes a novel inositol phospholipid, phospha-

tidylinositol-3-phosphate. Nature 1988; 332: 644-646.

31. Serunian LA, Haber MT, Fukui T et al. Polyphosphoinositides produced by the phosphatidylinositol 3-kinase are poor substrates for phospholipase C from rat liver and bovine brain. J Biol Chem 1989; 264: 17809-17815.
32. Otsu M, Hiles I, Gout I et al. Characterization of two 85kd proteins that associate with receptor tyrosine kinases, middle T/pp60$^{c\text{-}src}$ complexes, and PI3 kinase. Cell 1992; 65: 91-104.
33. Hiles I, Otsu M, Volinia S et al. Phosphatidylinositol-3-kinase: structure and expression of the 110 kDa catalytic subunit. Cell 1992; 70: 419-429.
34. Klippel A, Escobedo JA, Hu Q et al. A region of the 85-kilodalton subunit of phosphatidylinositol 3-kinase binds the 110-kDa catalytic subunit. Mol Cell Biol 1993; 13: 5560-5566.
35. Burr JG, Dreyfuss G, Penman S et al. Association of the *src* gene product of Rous sarcoma virus wirh cytoskeletal structures of chicken embryo fibroblasts. Proc Natl Acad Sci USA 1980; 77: 3484-3488.
36. Hamaguchi M, Hanafusa H. Association of p60^{src} with Triton X-100-resistant cellular structure correlates with morphological transformation. Proc Natl Acad Sci USA 1977; 84: 2312-2316.
37. Loeb DM, Woolford J, Beemon K. pp60$^{c\text{-}src}$ has less affinity for the detergent-insoluble cellular matrix than do pp60$^{v\text{-}src}$ and other viral protein-tyrosine kinases. J Virol 1987; 61: 2420-2427.
38. Ballmer-Hofer K, Ziegler A, Burger MM. Association of middle T antigen and pp60^{src} with cytoskeletal elements. Oncogene 1988; 3: 365-371.
39. Sadowski I, Stone JC, Pawson T. A noncatalytic domain conserved among cytoplasmic tyrosine kinases modifies the function and transforming ability of Fujinami sarcoma virus p130$^{gag\text{-}fps}$. Mol Cell Biol 1986; 6: 4396-4408.
40. Smith DB, Johson KS. Single-step purification of polypeptides expresed in Escherichia coli as fusions with glutathione-S-transferase. Gene 1988; 67:31-40.
41. Anderson D, Koch CA, Grey L et al. Binding of SH2 domains of phospholipase C gamma1, GAP and *src* to activated growth factor receptors. Science 1990; 250: 979-982.
42. Margolis B, Li N, Koch A et al. The tyrosine phosphorylated carboxylterminus of the EGF receptor is a binding site for GAP and PLC-gamma. EMBO J 1990; 9: 4375-4380.
43. Mayer BJ, Hanafusa H. Association of the *v-crk* oncogene product with phosphotyrosine-containing proteins and protein kinase activity. Proc Natl Acad Sci USA 1990; 87: 2838-2642.
44. Mayer BJ, Jackson PK, Baltimore D. The noncatalytic *src* homology region 2 of abl tyrosine kinase binds to tyrosine-phosphorylated cellular proteins with high affinity. Proc Natl Acad Sci USA 1991; 88: 627-631.
45. Moran MF, Koch CA, Anderson D et al. *src* homology 2 domains direct protein-protein interactions in signal transduction. Proc Natl Acad Sci USA 1990; 87: 8622-8626.
46. Cohen B, Yoakim M, Piwnica-Worms H et al. Tyrosine phosphorylation is a signal for traffiking of pp85, a polypeptide associated with phosphatidylinositol kinase activity. Proc Natl Acad Sci USA 1990; 87: 4458-4462.
47. Kaslauskas A, Cooper JA. Autophosphorylation of the PDGF receptor in the kinase insert region regulates interactions with cell proteins. Cell 1989; 58: 1121-1133.
48. Talmage DA, Freund R, Young AT et al. Phosphorylation of middle T by pp60$^{c\text{-}src}$: a switch for binding phosphatidylinositol 3-kinase and optimal tumorigenesis.Cell 1989; 59: 55-65.
49. Whitman M, Kaplan DR, Schaffhausen B et al. Association of phosphatidylinositol kinase activity with middle-T competent for transformation. Nature (London) 1985; 239-242.
50. Escobedo JA, Kaplan DR, Kavanaugh WM et al. A phosphatidylinositol 3-kinase binds to platelet-derived growth factor receptors through a specific receptor sequence containing phosphotyrosine. Mol Cell Biol 1991; 11: 1125-1132.

51. Fantl WJ, Escobedo JA, Martin GA et al. Distinct phosphotyrosines on a growth factor receptor bind to specific molecules that mediate different signaling pathways. Cell 1992; 69: 413-423.
52. Songyang Z, Shoelson SE, Chaudhuri M et al. SH2 domains recognize specific phosphopeptide sequences. Cell 1992; 72: 767-778.
53. Felder S, Zhou M, Hu P et al. SH2 domains exhibit high-affinity binding to tyrosine-phosphorylated peptides yet also exhibit rapid dissociation and exchage. Mol Cell Biol 1993; 13: 1449-1455.
54. Panayotou G, Gish G, End P et al. Interactions between SH2 domains and tyrosine-phosphorylated platelet-derived growth factor b-receptor sequences: analysis of kinetic parameters by a novel biosensor approach. Mol Cell Biol 1993; 13: 3567-3576.
55. Waksman G, Kominos D, Robertson SR et al. Crystal structure of the phosphotyrosine recogntition domain SH2 of *v-src* complexed with tyrosine-phosphorylated peptides. Nature (London) 1992; 358:646-653.
56. Overduin M, Mayer b, Rios CB et al. Scendary structure of the *src* homology 2 domain of *c-abl* by heteronuclear NMR spectroscopy in solution. Proc Natl Acad Sci USA 1992; 89: 11673-11677.
57. Overduin M, Rios CB, Mayer CB et al. Three-dimensional solution structure of the *src* homology 2 domain of *c-abl*. Cell 1992; 70: 697-704.
58. Waksman G, Shoelson SE, Pant N et al. Binding of a high affinity phosphotyrosyl peptide to the *src* SH2 domain: crystal structures of the complexed and peptide-free forms. Cell 1993; 72: 779-790.
59. Pawson T, Gish G. SH2 and SH3 domains: from structure to function. Cell 1992; 71: 359-362.
60. Birge RB, Hanafusa H. Closing in on SH2 specificity. Science 1993; 262: 1522-1524.
61. Panayotou G, Bax B, Gout I et al. Interaction of the p85 subunitn of PI 3-kinase and its N-terminal SH2 domain with a PDGF receptor phosphorylation site: structural features and analysis of conformational changes. EMBO J 1992; 11: 4261-4272.
62. Shoelson SE, Sivaraj M, Williams KP et al. Specific phosphopeptide binding regualtes a conformational change in the PI3 kinase SH2 domain associated with enzyme activation. EMBO J 1993;
63. Booker G, Breeze AL, Downing AK et al. Structure of an SH2 domain of the p85 alpha subunit of PI-OH kinase. Nature (London) 1992; 358: 684-687.
64. Mori SM, Ronnstrand L, Zilberstein A et al. Identification of two juxtamembrane autophosphorylation sites in the PDGF β-receptor: involvement in the interaction with *src* family tyrosine kinases. EMBO J 1993; 12:2257-2264.
65. Kashishian A, Kaslauskas A, Cooper JA. Phosphorylation sites in the PDGF receptor with different specificities for binding GAP and PI3 kinase in vivo. EMBO J 1992; 11: 1373-1382.
66. Kaslauskas A, Kashishian A, Cooper JA et al. GTPase-activating protein and phosphatidylinositol 3-kinase bind to distinct regions of the platelet-derived growth factor receptor b subunit. Mol Cell Biol 1992; 12: 2534-2544.
67. Nishimura R, Li W, Kashishian A et al. Two signaling molecules share a phosphotyrosine binding site in the PDGF receptor. Mol Cell Biol 1993; 13: 6889-6896.
68. Escobedo JA, Williams LT. A PDGF receptor domain essential for mitogenesis but not for other responses to PDGF. Nature (London) 1988; 335: 85-87.
69. Ronnstrand L, Mori S, Arridson AK et al. Identification of two C-terminal autophosphorylation sites in the PDGF β receptor: involvement in the interaction with phospholipase Cγ. EMBO J 1992; 11: 3911-3919.
70. Valius M, Bazenet C, Kazlauskas A. Tyrosines 1021 and 1009 are phosphorylation sites in the carboxy terminus of the platelet-derived growth factor receptor β subunit and are required for the binding of phospholipase Cγ and a 64-kilodalton protein, respectively. Mol Cell Biol 1993; 13: 133-143.
71. Kazlauskas A, Feng G-S, Pawson T et al. The 64-kilodalton protein that associates with the PDGF receptor subunit via tyrosine 1009 is the SH2 containing phosphotyrosine

phosphatase syp/SH-PTP2/PTP1D/SH-PTP3. Proc Natl Acad Sci USA 1993; 90: 6939-6943.

72. Valius M, Kazlauskas A. Phospholipase C-γ1 and phosphatidylinositol 3 kinase are the downstream mediators of the PDGF receptor's mitogenic signal. Cell 1993; 321-334.
73. Peters KG, Marie J, Wilson E et al. Point mutataion of an FGF receptor abolishes phosphatidylinositol turnover and Ca^{2+} flux but not mitogenesis. Nature (London) 1992; 358: 678-681.
74. Mohammedi M, Dionne CA, Li W et al. Point mutation in FGF receptor eliminates phosphatidylinositol hydrolysis without affecting mitogenesis. Nature (London) 1992; 358:681-684.
75. Panayotou G, Waterfield MD. Phosphatidylinositol 3-kinase: a key enzyme in diverse signaling processes. Trends Cell Biol 1992; 2: 358-360.
76. Corey S, Eguinoa A, Puyana-Theall K et al. Granulocyte-macrophage colony stimulating factor stimulates both association and activation of phosphoinositide 3OH-kinase and *src*-related tyrosine kinases(s) in human myeloid derived cells. EMBO J 1993; 12: 2681-2690.
77. Fox JEB, Lipfert L, Clark EA et al. On the role of platelet membrane skeleton in existing signal transduction. J Biol Chem 1993; 268: 25973-25984.
78. Zhang J, Fry MJ, Waterfield MD et al. Activated phosphoinositide 3-kinase associates with the membrane skeleton in thrombin-activated platelets. J Biol Chem 1992; 267: 4686-4692.
79. Grondin P, Plantavid M, Sultan C et al. Interaction of $pp60^{c\text{-}src}$, phospholipase C, inositol-lipid and diacylglycerol kinases with the cytoskeletons of thrombin-stimulated platelets. J Biol Chem 1991; 24: 15705-15709.
80. Yano H, Nakanishi S, Kimura et al. Inhibition of histamine secretion by wortmannin through the blockade of phosphatidylinositol 3-kinase in RBL-2H3 cells. J Biol Chem 1993: 268: 25846-25856.
81. Arcaro A, Wymann MP. Wortmannin is a potent phosphatidylinositol 3-kinase inhibitor: the role of phosphatidylinositol 3,4,5-trisphosphate in neutrophil responses. Biochem J 1993; 296: 297-301.
82. Wymann M, Arcaro A. Platelet-derived growth factor-induced phosphatidylinositol 3-kinase activation mediates actin rearrangements in fibroblasts. Biochem J 1994; 298: 517-520.
83. Parsons JT, Weber MJ. Genetics of *src*: structure and functional organization of a protein tyrosine kinase. Curr Top Microbiol Immunol 1989; 147: 80-127.
84. Bagrodia S, Chackalaparampil I, Kmiecik TE et al. Altered tyrosine 527 phosphorylation and mitotic activation of p60*c-src*. Nature (London) 1991; 349: 172-175.
85. Schuh SM, Brugge JS. Investigation of factors that influence phosphorylation of $pp60^{c\text{-}src}$ on tyrosine 527. Mol Cell Biol 1988; 8: 2465-2471.
86. Cooper JA, Gould KL, Cartwright CA et al. Tyr 527 is phosphorylated in $pp60^{c\text{-}src}$: implications for regulation. Science 1986; 231: 1431-1434.
87. Cooper JA, King CS. Dephosphorylation or antibody binding to the carboxy terminus stimulates $pp60^{c\text{-}src}$. Mol Cell Biol 1986; 6: 4467-4477.
88. Piwnica-Worms H, Suanders KB, Roberts TM et al. Tyrosine phosphorylation regulates the biochemical and biological properties of $pp60^{c\text{-}src}$. Cell 1987; 49: 75-82.
89. Reynolds AB, Vila J, Lansing TJ et al. Activation of the oncogenic potential of the avian cellular *src* protein by specific structural alteration of the carboxy terminus. EMBO J 1987; 6: 2359-2364.
90. Smart JE, Opperman H, Czernilofsky AP et al. Characterization of sites for tyrosine phosphorylation in the transforming protein of Rous sarcima virus ($pp60^{v\text{-}src}$) and its normal cellular homolog ($pp60^{c\text{-}src}$). Proc Natl Acad Sci USA 1981; 78:6013-6017.
91. Cooper JA, Howell B. The when and how of *src* regulation. Cell 1993; 73: 1051-1054.
92. Roussel RR, Brodeur SR, Shalloway D et al. Selective binding of activated $pp60^{c\text{-}src}$ by an immobilized phosphopeptide modeled on the carboxyl terminus of $pp60^{c\text{-}src}$.

Proc Natl Acad Sci USA 1991; 88: 10696-10700.

93. Sieh M, Bolen JB, Weiss A. CD45 specifically modulates binding of lck to a phosphopeptide encompassing the regulatory tyrosine of lck. EMBO J 1993; 12: 315-321.
94. Sabe H, Okada M, Nakagawa H et al. Activation of c-src in cells bearing v-crk and its expression by csk. Mol Cell Biol 1992; 12: 4706-4713.
95. Imamoto A, Soriano P. Disruption of the csk gene, encoding a negative regulator of the src family of tyrosine kinase, leads to neural tube defects and embryonic lethality in mice. Cell 1993; 73: 1117-1124.
96. Nada S, Yagi T, Takeda H et al. Constitutive activation of src family kinases in mouse embryos that lack csk. Cell 1993; 73: 1125-1135.
97. Drubin DJ, Mulholland J, Zhu Z et al. Homology of a yeast actin-binding protein to signal transduction proteins and myosin 1. Nature (London) 1990; 343: 288-290.
98. Koch CA, Anderson D, Moran MF. SH2 and SH3 domains: elements that control interactions of cytoplasmic signaling proteins. Science 1991; 252: 668-674.
99. Bar-Sagi D, Rotin D, Mandiyan V et al. SH3 domains direct cellular localization of signaling molecules. Cell 1993; 74: 83-91.
100. Mayer BJ, Baltimore D. Signaling through SH2 and SH3 domains. Trends Cell Biol 1993; 3: 8-13.
101. Cicchetti P, Mayer BJ, Theil G et al. Identification of a protein that binds to the SH3 region of abl and is similar to Bcr and GAP-rho. Science 1992; 257: 803-806.
102. Ren R, Mayer BJ, Cicchettt P et al. Identification of a ten-amino acid proline-rich SH3 binding site. Science 1993; 259: 1157-1161.
103. Rosakis-Adcock M, Fernley R, Wade S et al. The SH2 and SH3 domains of mammalian Grb2 couple the EGF-receptor to the ras activator mSos1. Nature (London) 1993; 363: 83-85.
104. Gout I, Dhand R, Hiles I et al. The GTPase dynamin binds to and is activated by a subset of SH3 domains. Cell 1993; 75: 265-36.
105. Herskovits JS, Shpetner HS, Burgess CC et al. Microtubules and scr homology 3 domains stimulate the dynamin GTPase via its C-terminal domain. Proc Natl Acad Sci USA 1993; 90: 11468-11472.
106. Herskovits JS Burgess CC, Obar RA et al. Effects of mutant rat dynamin on endocytosis. J Cell Biol 1993;122: 565-578.
107. Weng Z, Taylor JA, Turner CE et al. Detection of *src* homology 3-binding proteins, including paxillin, in normal and *v-src* transformed Balb/c 3T3 fibroblasts. J Biol Chem 1993; 268: 14956-14963.
108. Lui X, Marengere LE, Koch CA et al. The *v-src* SH3 domain binds phosphatidylinositol kinase. Mol Cell Biol 1993; 13: 5225-5232.
109. Barfod ET, Zheng Y, Kuang WJ et al. Cloning and expression of a human CDC42 GTPase-activating protein reveals a functional SH3-binding domain. J Biol Chem 1993; 268: 26059-26062.
110. Yu H, Chen JK, Feng S et al. Structural basis for the binding of proline-rich peptides to SH3 domains. Cell 1994; 76: 933-946.
111. Okamura N, Curnutte JT, Roberts RL et al. Relationship of protein phosphorylation to the activation of the respiratory burst in human neutrophils: defects in the phosphorylation of a group of closely related 48-kilodalton proteins in teo forms of chronicv granulomatous disease. J Biol Chem 1988; 263: 6777-6782.
112. Nauseef WM, Volpp BD, Leidal KG et al. Assembly of the neutrophil respoiratory burst oxides. Protein kinase C promotes cytoskeletal and membrane association of cytosolic oxidase components. J Biol Chem 1991; 266: 5911-5917.
113. Rotrosen D, Leto TL. Phosphorylation of neutrophil 47-kDa cytosolic oxidase factor. Translocation to membrane is associates with distinct phosphorylation events. J Biol Chem 1990; 19910-19915.
114. Pleiman CP, Hertz WM, Cambier JC. Activation of phosphatidylinositol-3' kinase by *src*-family kinase SH3 binding to p85 subunit. Science 1994; 263: 1609-1612.
115. Yu H, Rosen MK, Shin TB et al. Solution

structure of the SH3 domain of *src* and identification of its ligand-binding site. Science 1992; 258: 1665-1668.
116. Noble MEM, Mussacchio A, Saraste M et al. Crystal structure of the SH3 domain in human fyn: comparison of the three-dimensional structures of SH3 domains in tyrosine kinases and spectrin. EMBO J 1993; 12: 2617-2624.
117. Kohda D, Hatanaka H, Okada M et al. Solution structure of the SH3 domain of phospholipase C-γ. Cell 1993; 72: 953-960.
118. Koyama S, Yu H, Dalgarno DC et al. Structure of the PI3K SH3 domain and analysis of the SH3 family. Cell 1993; 72: 945-952.
119. Yu H, Chen J, Feng S et al. Structural basis for the binding of proline-rich peptides to SH3 domains. Cell1994; 76: 933-945.
120. Yang YS, Garbay C, Duchesne M et al. Solution structure of GAP SH3 domain by 1H NMR and spatial arrangement of esential Ras signaling-involved sequence. EMBO J 1994; 13: 1270-1279.
121. Simon MA, Bowtell BTL, Dodson GS et al. Ras1 and a putative guanine nucleotide exchange factor perform crucial steps in signaling by sevenless proetin tyrosine kinse. Cell 1991; 67: 701-716.
122. Bonfini L, Karlovich CA, Dasgupta C et al. The son of sevenless gene product: A putative activator of ras. Proc Natl Acad Sci USA 1992; 25: 603-606.
123. Bowtell D, Fu P, Simon M et al. Identification of murine homologues of the drosophila son of sevenless gene: potential activators of ras. Proc Natl Acad Sci USA 1992; 89: 6511-6515.
124. Chardin P, Camonis J, Gale WL et al. Human Sos1: a guanine nucleotide exchange factor for ras that binds to Grb2. Science 1993; 260: 1338-1343.
125. Clark SG, Stern MJ, Horvitz HR. C elegans cell-signaling gene Sem-5 encodes a protein with SH2 and SH3 domains. Nature (London) 1992; 356: 340-344.
126. Simon MA, Dodson GS, Rubin GM. An SH3-SH2-SH3 protein is required for $p21^{ras}$ activation and binds to Sevenless and Sos proteins in vitro. Cell 1993; 73: 169-177.
127. Olivier JP, Raabe T, Henderson M et al. A drosophila SH2-SH3 adaptor protein implicated in coupling the sevenless tyrosine kinase to an activator of ras guanine nucleotide exchange. Cell 1993; 73: 179-191.
128. Lowenstein EJ, Daly RJ, Batzer AG et al. The SH2 and SH3-containing protein GRB2 links receptor tyrosine kinases to ras signaling. Cell 1992; 70: 131-142.
129. Gale WN, Kaplan D, Lowenstein EJ et al. Grb2 mediates the EGF-dependent activation of guanine nucleotide exchange on ras. Nature (London) 1993; 363: 88-92.
130. Schlessinger J. SH2/SH3 signaling proteins. Curr Opinion Genet Dev 1994; 4: 25-30.
131. Li N, Batzer A, Daly R et al. Guanine nucleotide releasing factor hSos1 binds to Grb2 and links receptor tyrosine kinases to Ras signaling. Nature (London) 1993; 363: 85-88.
132. Buday L, Downward J. Epidermal growth factor regulates p21ras through the formation of a complex of receptor, Grb2 adaptor protein, and Sos nucleotide exchange factor. Cell 1993; 73: 611-620.
133. Egan SE, Giddings BW, Brooks MW et al. Association of Sos ras exchange protein with Grb2 is implicated in tyrosine kinase signal transduction and transformation. Nature (London) 1993; 363: 45-51.
134. Pelicci G, Lanfrnacona I, Grignani F et al. A novel transforming protein (SHC) with an SH2 domain is implicated in mitogenic signal transduction. Cell 1992; 70: 93-104.
135. Rozakis-Adcock M, McGlade J, Mbamalu G et al. Association of Shc and GRB2/sem-5 SH2 containing protein is implicated in activation of ras pathway by tyrosine kinases. Nature (London) 1992; 360: 689-692.
136. Skolnik EY, Lee CH, Batzer A et al. The SH2/SH3 domain containing protein GRB2 interacts with tyrosine phosphorylated IRS-1 and Shc: Implications for insulin control of ras signaling. EMBO J 1993; 12: 1929-1936.
137. Pronk GJ, McGlade J, Pelicci G et al. Insulin-induced phosphorylation of the 46- and 52 kDa Shc proteins. J Biol Chem 1993; 268:5748-5753.

138. Skolnik EY, Batzer A, Li N et al. Function of Grb2 in linking the insulin receptor to Ras signaling pathway. Science 1993; 260: 1953-1955.
139. Matuoka K, Shibasaki F, Shibata M et al. Ash/Grb2, a SH2/SH3-containing protein, couples to signaling for mitogenesis and cytoskeletal reorganization by EGF and PDGF. EMBO J 1993; 12:3467-3473.
140. Miki H, Miura K, Matuoka K et al. Association of Ash/Grb-2 with dynamin through the *src* homology 3 domain. J Biol Chem 1994; 269: 5489-5492.
141. Trowbridge IS. Dynamin, SH3 domain and endocytosis. Current Biol 1993; 11:773-775.

CHAPTER 6

Tyrosine Kinases and Human Disease

The role of tyrosine kinases in human diseases such as cancer is currently being intensely investigated in laboratories throughout the world. Most of the investigations are focused on the role of growth factors and growth factor receptors since the pioneering work from the laboratories of Waterfield, Schlessinger, Doolittle and Bishop established that several viral encoded oncogenes were homologs of either growth factors such as PDGF (*v-sis*) or growth factor receptors such as EGF-R (*v-erbB*).[1-7] There are some clear-cut examples of association of tyrosine kinases with cancers which I will briefly discuss first, however since this book has concentrated on the structure and function of *src* most of this chapter will be spent on reviewing the possible role of the *src* gene in human cancers.

CHRONIC MYELOID LEUKEMIA

Chronic myeloid leukemia (CML) is characterized by a reciprocal chromosomal translocation t(9;22)(q34;q11) to produce the Philadelphia chromosome which is used diagnostically and is found in over 90% of cases of CML.[8,9] This translocation is also found in 20-50% of adults with acute lymphoblastic leukemia (ALL).[10,11] The result of this genetic rearrangement is to relocate the *abl* protooncogene to the breakpoint cluster region (bcr) on chromosome 22 to generate a hybrid *bcr-abl* gene product.[9] Two alternative hybrid proteins are produced: $p210^{bcr\text{-}abl}$ and $p185^{bcr\text{-}abl}$ which are characteristic of CML and ALL respectively.[12-14] The bcr sequences specifically activate the tyrosine kinase activity and transforming potential of bcr-abl.[15,-17] The fusion of bcr sequences to abl also results in an altered distribution of the bcr-abl such that there is enhanced association to microfilament bundles,[16] and the localization of bcr-abl to the cytoskeleton of CML cells may be an important factor in leukemias. The production of $p210^{bcr\text{-}abl}$ appears to result in the abnormal expansion of the myeloid progenitor pool in bone marrow and peripheral blood.[18] Although the biochemical mechanism by which CML cells are generated by chromosomal translocation is unknown there is now some evidence that the $p210^{bcr\text{-}abl}$ signals through to *ras* activation via SH2 and SH3 domains in a manner similar to that described in chapter 5 for other tyrosine kinases and signaling molecules. Thus it is possible that the molecular lesion is a disruption of a normal tyrosine kinase mediated signaling process. Prendergast et al[19] have provided evidence that the bcr-abl-induced oncogenesis is mediated by direct interaction of a bcr sequence with the SH2 domain of Grb2, since deletion of this binding inhibits signaling through to *ras*. Furthermore the SH2 domain of *bcr-abl* itself binds to other tyrosine phosphorylated proteins within cells although at present none of these proteins have been identified.[20]

One exciting development is the potential for gene therapy for leukemias such as CML and ALL which have a defined genetic lesion. Szczylik et al[21] treated CML blast cells with bcr-abl antisense oligonucleotides complementary to two bcr-abl junctions. Cell growth and colony formation were suppressed concurrent with a reduction in the expression of *bcr-abl*, whereas granulocyte-macrophage colony formation from normal marrow was unaffected. These results demonstrate that a functional bcr-abl gene is necessary to maintain the leukemogenic cells, but also show that it is possible to selectively inhibit the proliferation of leukemic cells without affecting normal counterparts and offers the prospect of gene therapy for this disease.

SRC AND BREAST CANCER

Human breast carcinomas often overexpress the related receptors EGF receptor and $p185^{neu}$.[22] In addition amplification of these genes correlates with poor prognosis.[23,24,25] An indication that *c-src* might play a role in breast cancer comes from the observations many primary breast cancers show elevated $pp60^{c\text{-}src}$ activity. About 60 breast cancer specimens were screened by northern blotting for the expression of a number of protooncogenes and 25-30% were found to overexpress *c-ErbB*, *raf-1*, *src*, *lck* or *H-ras*. Somewhat surprisingly only 20% of this sample displayed overexpression of *neu*.[26] In a large study of 72 breast cancer specimens all of these tumors displayed elevated tyrosine kinase activity and at least 70% of this activity could be attributed to $pp60^{c\text{-}src}$,[27,28] so this protooncogene is at least implicated in breast cancer development.

Cells which overexpress $pp60^{c\text{-}src}$ show no overt phenotypic changes but hyperrespond to epidermal growth factor.[29] Moreover in order to induce this enhanced response the overexpressed $pp60^{c\text{-}src}$ must be functional, localized to the plasma membrane and contain an intact SH2 domain.[30] This hypothesis has been strengthened by the recent report that $pp60^{c\text{-}src}$ is involved in two signaling pathways in human breast cancer. Luttrell et al[31] examined the interactions of the SH2 and SH3 domains of *c-src* with proteins from human carcinoma cell lines and found that the *c-src* SH2 domain bound to activated epidermal growth factor receptor and $p185^{neu}$, and endogenous $pp60^{c\text{-}src}$ also associated with tyrosine-phosphorylated epidermal growth factor receptor. The association was inhibited by phosphopeptides homologous to the region surrounding tyrosine-530, the regulatory region of *c-src* implying that *c-src* associated with epidermal growth factor receptor via a carboxy terminal region. These workers proposed that the association of $pp60^{c\text{-}src}$ with epidermal growth factor or $p185^{neu}$ contributes to the unregulated proliferation and malignant transformation. Consistent with this suggestion are the findings that $pp60^{c\text{-}src}$ is hyperphosphorylated in M phase of the cell cycle, possibly mediated by *cdc2* which is activated by dephosphorylation at the G2/M boundary. It may be that in malignant cells $pp60^{c\text{-}src}$ is constitutively phosphorylated throughout the cell cycle leading to constitutive association with epidermal growth factor receptor and hyperresponsiveness to epidermal growth factor.[32-34] Breast cancer cells commonly proliferate in response to estrogens, and Migliaccio et al[35] have obtained results which might link this to tyrosine phosphorylation. Estrodiol stimulated the tyrosine phosphorylation of human breast cancer MCF-7 cells under conditions which also resulted in the stimulation of estrogen-dependent proliferation. Two major substrates were found at 55 kDa and

60 kDa and these were identified as pp60$^{c\text{-}src}$ by polyclonal and monoclonal antibodies. The tyrosine phosphorylation could be inhibited by the antiestrogens tamoxifen and ICI 164,384 suggesting that one of the signaling pathways of estrogens is to transiently stimulate tyrosine phosphorylation and activation of pp60$^{c\text{-}src}$ in addition to affecting gene expression in the nucleus. It is likely that other tyrosine kinases are also activated in breast cancer cells and recently 25 different protein kinases were isolated from a primary breast carcinoma and a breast cancer cell line of which three were novel and the others belonging to defined protein kinase classes such as JAK and *CDC2*.[36] Other potential signaling proteins which may interact with tyrosine kinases in breast cancer cells have been identified. A newly cloned protein which contains an SH2 domain, GRB-7, has been mapped to a region on mouse chromosome 11 which also contains *c-erbB2*. In man this locus is often amplified in breast cancer leading to overexpression of *c-erbB2*. Stein et al[37] have found that GRB-7 is overexpressed in several breast cancer cell lines and breast tumors. In addition these workers found that GRB-7 binds to *c-erbB2* and also to SHC. GRB-7 may play a role analogous to GRB-2 which binds to SHC and SOS and is central to the signaling pathways of the platelet-derived growth factor receptor and *v-src* transformation in fibroblasts (see chapter 5).

SRC AND NEUROBLASTOMA

Neuroblastoma is the most common malignant solid cancer of children and appears in different differentiation states, which has prognostic value. Neural tissues, including brain, retina, and spinal ganglia were shown to contain high levels of pp60$^{c\text{-}src}$-associated tyrosine kinase activity and express high levels of pp60$^{c\text{-}src}$ proteins, averaging 8- to 10-fold higher than in other tissues.[37-41] In the nervous system the expression of pp60$^{c\text{-}src}$ is developmentally regulated with high levels found in developing neural tissues and postmitotic, fully differentiated neurons.[42,43] Immunohistochemistry has localized pp60$^{c\text{-}src}$ within processes and cell bodies of ganglion neurons and the processes of rods and cones, and this was confirmed by tyrosine kinase assays.[41,44,45,46] In addition a novel *src* gene product initially detected as a structural alteration causing a retardation of the mobility of pp60$^{c\text{-}src}$ in SDS gels,[47] pp60$^{c\text{-}srcN1}$, has been identified only in neurons and related tissues. This protein has an extra six amino acids following residue 114 in the SH2 domain derived from 18 additional bases transcribed by alternative splicing of the intron between exons 3 and 4.[48-52] It has been reported that this larger form has higher tyrosine kinase activity and two or more additional sites of serine phosphorylation within the amino-terminal 18 kDa[39,47] pp60$^{c\text{-}srcN1}$ has a more limited tissue distribution and is predominant in the mature central nervous system whereas pp60$^{c\text{-}src}$ is expressed in both the central and peripheral nervous systems. Another larger pp60$^{c\text{-}src}$ species with an additional 33 bases capable of coding 11 extra amino acids has been cloned from human brain. Analysis of brain tissue mRNA revealed that these additional bases were utilized primarily in conjunction with the other insert described earlier to yield a product possessing 17 additional amino acids, pp60$^{c\text{-}srcN2}$. The function of this differentially spliced product has yet to be investigated,[53] although the insert is in the SH3 domain which might alter the ability of this protein to interact with other signaling molecules. The presence of *c-src*, *c-srcN1* and *c-srcN2* mRNA were examined in relation to neural differentiation in eight neuroblastoma cell lines and seven primary neuroblastomas.[54,55] *c-srcN1* was expressed in cell lines with the ability to differentiate but not in cells lacking this response and *c-srcN2* increased in differentiated neuroblastomas after induction. Infantile neuroblastomas expressed both *c-srcN1* and *c-srcN2* whereas neuroblastomas from older children or differentiated neuroblastomas did not. These results suggest that there is a correlation between *c-srcN* expression and the differentiation state of neuroblastomas. Since the *src* gene product is highly expressed in such non-dividing cells it has been suggested that the src gene in developing neural tissue is involved in differentiation rather than proliferation, and the effects of *v-src* on neural cell differentiation were discussed in chapter 2. Neuronal differentiation of mouse embryo carcinoma cells or differentiation of rat neuroblasts result in an increase in pp60$^{c\text{-}src}$ protein and kinase activities indicating a close link between this protein and the neuronal phenotype.[39,51] One of the first hints that pp60$^{c\text{-}src}$ might be involved in deregulation of neuronal cell growth came from the work of Bolen et al[56] who compared the levels of pp60$^{c\text{-}src}$ and its kinase activity in human neuroblastomas and glioblastomas with that of fibroblasts. These workers found that although pp60$^{c\text{-}src}$ expression was not significantly increased in four neuroblastoma cell lines when protein or mRNA were measured there was a 20- to 40-fold increase in the pp60$^{c\text{-}src}$-associated tyrosine kinase activities compared to four glioblastoma cell lines or normal human fibroblasts. Moreover the *c-src* protein from neuroblastomas ran more slowly on SDS gels and contained a unique amino-terminal tyrosine phosphorylation site. To examine more closely the link between pp60$^{c\text{-}src}$ activity and neuroblastoma function the levels of *src* protein and kinase activity were measured in two different pairs of genetically indistinguishable cloned neuroblastoma cell variants which have different phenotypes. Cells which possessed the neuroblastic phenotype contained high levels of pp60$^{c\text{-}src}$ protein whereas cells which displayed the substrate-adherent phenotype had low pp60$^{c\text{-}src}$ levels.[56] The increased enzymatic activity of neuronal c-src is still unclear since these workers found no difference in the tyrosine kinase activity of the pp60^{src} proteins expressed by the different cell lines. In contrast if the SH-SY5Y neuroblastoma cell line was induced to differentiate down the neuronal pathway the levels of *src* protein were unaffected but there was a threefold increase in enzyme specific activity.[57] In this same cell line a complex between pp60$^{c\text{-}src}$/pp60$^{c\text{-}srcN}$ and a 38 kDa protein was found in the growth cones after differentiation,[58] and protein and kinase activity are highly enriched in growth cone fractions of embryonic rat brain.[59]

The role of pp60$^{c\text{-}src}$/pp60^{srcN} in neuroblastomas or neuronal cells is still virtually unknown, although several substrates for its tyrosine kinase activity have been identified which might affect function as described in chapter 4. In particular tubulin in growth cones is tyrosine phosphorylated resulting in a reduced capacity to polymerise which might lead to phenotypic changes involving motility or proliferation and a synaptic vesicle protein synaptophysin is also tyrosine phosphorylated which might lead to defects in intercellular communication.

SRC AND COLON CARCINOMA

Colon carcinoma is one of the most common human malignancies and so any clues as to the factors involved in the development of this neoplasia are important. A series of reports have indicated that there may be a causative relationship between colon carcinoma and the activation of pp60$^{c\text{-}src}$.

Bolen et al[60] examined pp60$^{c\text{-}src}$ kinase activity in 21 carcinoma cell lines and 5 normal colon mucosal cell lines and found that all the neoplastic lines had an elevated kinase activity compared to normal cells, averaging at an 80-fold difference in autophosphorylation. These workers then measured the activity of pp60$^{c\text{-}src}$ derived directly from 15 colon adenocarcinoma tissues and showed that all the tumor tissues displayed elevated activity compared to adjacent normal tissue. However the pp60$^{c\text{-}src}$ protein level in the tumor tissue did not directly correlate with activity levels, suggesting an elevation in the pp60$^{c\text{-}src}$ kinase activity rather than increased gene expression. It was later confirmed that levels of pp60$^{c\text{-}src}$ mRNA were similar in normal and tumor cell lines. The biochemical mechanisms for the increased *src* activity are still not completely understood, since there were no detectable alterations in the *src* phosphorylation sites in normal mucosal or carcinoma cells, however there was increased phosphate turnover in the neoplastic cells which could contribute to functional *src* changes.[61] Further studies showed no detectable differences in the phosphotyrosyl phosphatase activity between the normal and cancer cells, so the mechanism of the increased phosphate turnover could not be attributed to phosphatase activity. On the other hand if phosphotyrosine phosphatase activity was inhibited by vanadate there was preferential stimulation of tyrosine kinase activity in the colonic carcinoma cells rather than the normal colonic cells, consistent with the earlier investigations suggesting differential pp60$^{c\text{-}src}$ regulation.[62]

Similar results were obtained by Takekura et al[64] who found that 50% of gastric carcinomas tested contained higher pp60$^{c\text{-}src}$ kinase activity than the corresponding normal mucosa. Cartwright et al[65] studied the development of adenoma polyps from normal epithelia with respect to tyrosine kinase activity and found that the polyps also contained elevated tyrosine kinase activity. These studies also showed that activity was highest in malignant polyps and in large benign polyps which demonstrated severe dysplasia, indicating that pp60$^{c\text{-}src}$ activation occurs in benign polyps which are at greatest risk for developing cancer. The investigations on the role of tyrosine kinase activity in colon carcinoma were extended by Cartwright et al[65] who found that many colon carcinoma cell lines contained higher pp60$^{c\text{-}src}$ kinase activities than corresponding normal tissue. Immunoblotting of colon carcinoma tissue revealed that three phosphotyrosine-containing proteins of 145 kDa, 125 kDa and 57 kDa were prominent in eight colon carcinoma cell lines but at lower levels in normal mucosa cells. These proteins are as yet unidentified but all the cell lines which exhibited tyrosine phosphorylation of these three proteins also displayed elevated pp60$^{c\text{-}src}$-associated tyrosine kinase activity. Thus it seems likely that elevated pp60$^{c\text{-}src}$ kinase activity contributes to altered protein phosphotyrosine patterns in colon carcinoma cells. Another protein of 63 kDa has been found to have an altered tyrosine phosphorylation pattern in colon carcinoma cells. In this case changes in tyrosine phosphorylation occurred in the cancer cells after induction of mitogenesis by diglycerides, which did not induce proliferation of normal colonic cells.[66] Whilst several other proteins were phosphorylated on tyrosine in some of the carcinoma cell, there was diglyceride-induced tyrosine phosphorylation of the 63 kDa protein in all the carcinoma cells tested. It is possible that this is a functional substrate important for colon tumorigenesis. There may also be differences in the pp60$^{c\text{-}src}$ tyrosine kinase activity of colon carcinomas in different metastatic site.[67]

In a study comparing the tyrosine kinase of pp60$^{c\text{-}src}$ from a variety of colon carcinomas Weber et al[69] found that there was a good correlation between the extent of carcinoma differentiation and kinase activity, with high activity in well differentiated tumor material and low activity in poorly differentiated material. This suggests a role for pp60$^{c\text{-}src}$ in the colon differentiation, and this is consistent with the observations that activation of the tyrosine kinase activity of pp60$^{c\text{-}src}$ is an early event in colon polyps development,[64,68] and may precede polyp or tumor formation.[69] This activation may well be related to the differentiation state of the cell since there is some evidence of differential protein kinase distribution and different tyrosine kinase activities in normal colon cell differentiation and/or proliferation.[70] It is also interesting to note that intestinal crypt stem cells possess high levels of cytoskeletal associated tyrosine phosphorylated proteins and tyrosine kinase activity compared to undifferentiated enterocytes.[71]

The role of protooncogenes and oncogene cooperation in the etiology of colonic cancer has also been investigated. D'Emilia et al[73] introduced viral vectors harboring *myc* or *src* oncogenes into Wistar rats. *Myc* alone induced atypia and some dysplasia whereas src alone induced focal dysplastic lesions and metaplasia throughout the colon. In contrast *myc* and *src* introduced together produced highly dysplastic lesions and tumor formation at later time points. These results indicate that src may be involved in the development of colonic carcinoma, however cooperation between oncogenes is an important factor for full tumor formation. Similar experiments have been performed in vitro. Single oncogenes such as *v-myc*, *v-src* or *v-H-ras* introduced into normal fetal rat colonic tissue in organ culture using retroviral vectors resulted in enhanced proliferation but no establishment of cell lines. Pairs of oncogenes such as *myc/ras* or *myc/src* resulted in cell lines in more than 80% of the experiments with different phenotypes depending on the oncogene pairs, again emphasizing the importance of oncogene cooperation in colon carcinoma.[73]

I have previously suggested that association of $pp60^{v\text{-}src}$ with the cytoskeleton is an important feature of neoplastic transformation. Most studies have failed to provide any evidence for the association of $pp60^{c\text{-}src}$ with the cytoskeleton, except in the case of platelets as described in chapters 3 and 4. Schwartz et al[75] investigated the tyrosine kinase activity in colonic epithelium and found that this activity was associated with cytoskeleton-enriched subcellular fractions. This cytoskeleton-associated kinase activity was enhanced in malignant carcinomas compared with uninvolved tissue. It is fascinating to suggest that this cytoskeleton-associated kinase activity is partly due to $pp60^{c\text{-}src}$, that this is upregulated in colonic cancer, and that functional substrates might be cytoskeletal components.

If the activation of a *src*-specific tyrosine kinase is an important feature of colon carcinomas, is it possible that tyrosine kinase inhibitors might be useful therapeutic drugs? Garcia et al[76] examined the effect of the tyrosine kinase inhibitor herbimycin A on the growth of seven colon tumor cell lines. These workers found that in careful dose-response studies all the tumor cell lines were more sensitive to herbimycin A than normal colon cells, and that herbimycin-induced inhibition of proliferation was reversible. These results hint that agents which inhibit tyrosine phosphorylation may efficacious in the therapy of colon carcinoma.

OTHER CANCERS

Although there is no widespread evidence, $pp60^{c\text{-}src}$ overexpression may be implicated in other cancers. Reported examples are small cell or non-small cell carcinomas[76] and adenoma of the esophagus.[77] Five- to ten-fold elevation of *c-yes*, but not *c-src* has been reported in human malignant melanoma.[78] Much more information is necessary before a general role for oncogenic tyrosine kinases can be established for tumors other than those already discussed.

Multiple endocrine neoplasia type 2A (MEN2A) is a dominantly inherited syndrome affecting tissues derived from neural ectoderm, characterized by medullary thyroid carcinoma and pheochromocytoma. Several groups have localized the genetic lesion to a region in chromosome 10q11.2 in virtually all MEN2A tumors examined.[80-84] This genetic locus contains the RET gene which encodes a tyrosine kinase, and MEN2A tumors have been shown to be mutated in the RET gene resulting in fusion to other genes, in some cases giving rise to a fusion protein of the RET tyrosine kinase and the regulatory subunit of protein kinase A. The resulting products can have different subcellular locations and tyrosine kinase activities.[85-89] The pathway leading from activated RET tyrosine kinase to tumor production is not yet known, but this tumor may be an exciting model for studying tyrosine kinase-induced neoplasia in humans.

References

1. Waterfield MD, Scrace GT, Whittle N et al. Platelet-derived growth factor is structurally related to the putative transforming protein $p28^{sis}$ of simian sarcoma virus. Nature (London) 1983; 304: 35-39.
2. Van Beveren C, Gallenshaw JA, Jonas V et al. Nucleotide sequence and formation of the transforming gene of mouse sarcoma virus. Nature (London) 1981; 289: 258-262.

3. Ullrich A, Schlessinger J. Signal transduction by receptors with tyrosine kinase activity. Cell 1990; 60: 755-765
4. Bishop JM. Molecular themes in oncogenesis. Cell 1991; 64: 235-248.
5. Downward J, Yarden Y, Mayes E et al. Close similarity of epidermal growth factor receptor and *v-erbB* oncogene protein sequences. Nature (London) 1984; 307: 521-527.
6. Cantley LC, Auger KR, Carpenter C et al. Oncogenes and signal transduction. Cell 1992; 64: 281-302.
7. Yarden Y, Ullrich A. Growth factor receptor tyrosine kinases. Ann Rev Biochem 1988; 57: 443-478.
8. Rowley, JD. A new consistent chromosomal abnormality in chronic myelogenous leukemia identified by quinacrine fluorescence and Giemsa staining. Nature (London) 1973; 243: 290-292.
9. Champlin RE, Golde DW. Chronic myelogenous leukemia: recent advances. Blood 1985; 65: 1039-1047.
10. Maurer J, Janssen JWG, Thiel E et al.Detection of chimeric BCR-ABL genes in acute lymphoblastic leukemia by the polymerase chain reaction. Lancet 1991; 337: 1055-1056.
11. Saglio G, Guerrasio A, Rosso C et al. Detection of Ph[1]-positive acute lymphoblastic leukemia by PCR. Lancet 1991; 338: 958-959.
12. Melo JV, Gordon DE, Cross NCP, Goldman JM. The ABL-BCR fusion gene is expressed in chronic myeloid leukemia. Blood 1993; 81: 158-165.
13. Walker LC, Ganesan TS, Dhut S et al. A novel chimeric protein is expressed in Philadelphia positive acute lymphoblastic leukemia. Nature (London) 1987; 329: 851-853.
14. Melo, JV, Gordon DE, Tuszynski A et al. Expression of the ABL-BCR fusion gene in Philadelphia-positive acute lymphoblastic leukemia. Blood 1993; 81: 2488-2491.
15. Muller AJ, Young JC, Prendergast AM et al. BCR first exon sequences specifically activate the BCR/ABL tyrosine kinase oncogene of Philadelphia chromosome-positive human leukemias. Mol Cell Biol 1991; 11: 1785-1792.
16. McWhirter JR, Wang JYJ. Activation of tyrosine kinase and microfilament-binding functions of *c-abl* by *bcr* sequences in *bcr/abl* fusion proteins. Mol Cell Biol 1991; 11: 1553-1565.
17. Lugo T, Prendergast AM, Muller AJ et al. Tyrosine kinase activity and transformation potency of BCR-ABL oncogene products. Science 1990; 247: 1079-1082.
18. Eaves AC, Cashman JD, Gaboury LA et al. Unregulated proliferation of primitive chronic myeloid leukemia progenitors in the presence of normal marrow adherent cells. Proc Natl Acad Sci USA 1986; 83: 5306-5310.
19. Prendergast AM, Quiliam LA, Cripe LD et al. BCR-ABL-induced oncogenesis is mediated by direct interaction with the SH2 domain of the GRB-2 adaptor protein. Cell 1993; 75: 175-185.
20. Mayer BJ, Jackson PK, Baltimore D. The noncatalytic *src* homology region 2 segment of abl tyrosine kinase binds to tyrosine-phosphorylated cellular proteins with high affinity. Proc Natl Acad Sci USA 1991; 88: 627-631.
21. Szczylik C, Skorski T, Nicoliades NC et al. Selective inhibition of leukemia cell proliferation by BCR-ABL antisense oligonucleotides. Science 1991; 253: 562-565.
22. Ulrich A, Schlessinger J. Signal transduction by receptors with tyrosine kinase activity. Cell 1990; 61: 203-212.
23. Slamon DJ, Clark GM, Wong SG et al. Human breast cancer: correlation of relapse and survival with amplification of the HER-2/*neu* oncogene. Science 1987; 235: 177-182.
24. Slamon DJ, Godolphin W, Jones LA et al. Studies of the HER-2/*neu* proto-oncogene in human breast and ovarian cancer. Science 1989; 244: 707-712.
25. Singletary SE, Baker FL, Spitzer G et al. Biological effect of epidermal growth factor on the in vitro growth of human tumors. Cancer Res 1987; 47: 403-406.
26. Koster A, Landgraf S, Leipold A et al. Expression of oncogenes in human breast cancer specimens. Anticancer Res 1991; 11: 193-201.
27. Ottenhoff-Kalff AE, Rijksen G, van Beurden EA et al. Characterization of protein tyrosine

kinases from human breast cancer: involvement of the *c-src* oncogene product. Cancer Res 1992; 52: 4773-4778.

28. Hennipman A, van-Oirschot BA, Smits J et al. Tyrosine kinase activity in breast cancer, benign breast disease, and normal breast tissue. Cancer Res 1989; 49: 516-521.
29. Luttrell LK, Luttrell LM, Parsons SJ. Augmented mitogenic responsiveness to epidermal growth factor in murine fibroblasts that overexpress $pp60^{c\text{-}src}$. Mol Cell Biol 1988; 8: 497-501.
30. Wilson LK, Luttrell DK, Parsons JT, Parsons SJ. $pp60^{c\text{-}src}$ tyrosine kinase, myristylation, and modulatory domains are required for enhanced mitogenic responsiveness to epidermal growth factor seen in cells overexpressing *c-src*. Mol Cell Biol 1989; 9: 1536-1544.
31. Luttrell DK, Lee A, Lansing TJ et al. Involvement of $pp60^{c\text{-}src}$ with two signaling pathways in human breast cancer. Proc Natl Acad Sci USA 1994; 91: 83-87.
32. Chackalaparampil I, Shalloway D. Altered phosphorylation and activation of $pp60^{c\text{-}src}$ during fibroblast mitosis. Cell 1988; 52: 801-810.
33. Morgan DO Kaplan JM Bishop JM et al. Mitosis-specific phosphorylation of $pp60^{c\text{-}src}$ by $p34^{cdc2}$-associated protein kinase. Cell 1989; 57: 775-786.
34. Shenoy S, Choi JK, Bagrodia S et al. Purified maturation promoting factor phosphorylates $pp60^{c\text{-}src}$ at the sites phosphorylated during fibroblast mitosis. Cell 1989; 57: 763-774.
35. Migliaccio A, Pagano M, Auricchio F. Immediate and transient stimulation of protein tyrosine phosphorylation by estrodiol in MCF-7 cells. Oncogene 1993; 8: 2183-2191.
36. Cance WG, Craven RJ, Weiner TM, Lui ET. Novel protein kinases expressed in human breast cancer. Int J Cancer 1993; 54: 571-577.
37. Cotton PC, Brugge JS. Neural tissues express high levels of the cellular *src* gene product $pp60^{c\text{-}src}$. Mol Cell Biol 1983; 3: 1157-1162.
38. Sein D, Wu J, Fuqua SAW et al. The SH2 domain protein GRB-7 is co-amplified, overexpressed and is in a tight complex with HER2 in breast cancer. EMBO J 1994; 13: 1331-1340
39. Brugge JS, Cotton PC, Queral AE et al. Neurones express high levels of a structurally activated form of $pp60^{c\text{-}src}$. Nature 1985; 316: 554-557.
40. Cartwright CA, Simantov R, Kaplan PL et al. Alterations in $pp60^{c\text{-}src}$ accompany differentiation of neurones from rat embryo striatum. Mol Cell Biol 1987; 7: 1830-1840.
41. Cartwright CA, Simantov R, Cowan WM et al. $pp60^{c\text{-}src}$ expression in the developing rat brain. Proc Natl Acad Sci USA 1988; 85; 3348-3352.
42. Sorge LK, Levy BT, Maness PF. $pp60^{c\text{-}src}$ is developmentally regulated in the neural retina. Cell 1984; 36: 249-257.
43. Levy BT, Sorge LK, Meymandi A et al. $pp60^{c\text{-}src}$ is in chick and human embryonic tissues. Dev Biol 1984; 104: 9-17.
44. Vardimon L, Fox LE, Moscona AA. Accumulation of c-src mRNA is developmentally regulated in embryonic neural retina. Mol Cell Biol 1986; 6: 4109-4111.
45. Fults DW, Towle AC, Lauder JM et al. $pp60^{c\text{-}src}$ in the developing cerebellum. Mol Cell Biol 1985; 5: 27-32.
46. Jacobs C, Rubsamen H. Expression of $pp60^{c\text{-}src}$ protein kinase in adult and fetal human tissue: high activities in some sarcomas and mammary carcinomas. Cancer Res 1983; 43: 1696-1702.
47. Sorge JP, Sorge LK, Maness PF. $pp60^{c\text{-}src}$ is expressed in human and fetal adult brain. Amer J Pathol 1985; 199: 151-157.
48. Brugge J, Cotton P, Lustig A et al. Characterization ofthe altered form of the *c-src* gene product in neuronal cells. Genes and Devel 1987; 1:287-296.
49. LeBeau JM, Wiestler OD, Walter G. An altered form of $pp60^{c\text{-}src}$ is expressed primarily in the central nervous system. Mol Cell Biol 1987; 7: 4115-4117.
50. Levy JB, Dorai LH, Wang L-H et el. The structurally distinct form of $pp60^{c\text{-}src}$ is detected in neuronal cells is encoded by a unique *c-src* mRNA. Mol Cell Biol 1987; 7: 4142-4245.
51. Martinez R, Mathey-Prevot B, Bernards A et al. Neuronal $pp60^{c\text{-}src}$ contains a six-amino

acid insertion relative to its non-neuronal counterpart. Science 1987; 237: 411-415.

52. Lynch SA, Brugge JS, Levine JM. Induction of altered *c-src* product during neural differentiation of embryonal carcinoma cells. Science 1986; 234: 8773-876.
53. Yang X, Martinez R, LeBeau J et al. Evolutionary expression of the neuronal form of the *src* protein in the brain. Proc Natl Acad Sci USA 1989; 86: 4751-4755.
54. Pyper JM, Bolen JB. Identification of a novel neuronal C-SRC exon expressed in human brain. Mol Cell Biol 1990; 10: 2035-2040.
55. Matsunaga T, Shirasawa H, Tanabe M et al. Expression of alternatively spliced *src* messenger RNAs related to neuronal differentiation in human neuroblastomas. Cancer Res 1993; 53: 3179-3185.
56. Bolen JB, Rosen N, Israel MA. Increased pp60$^{c\text{-}src}$ tyrosyl kinase activity in human neuroblastomas is associated with amino-terminal tyrosine phosphorylation of the *src* gene product.proc Natl Acad Sci USA 1985; 82: 7275-7279.
57. Veillette A, O'Shaughnessy J, Horak ID et al. Coordinate alteration of pp60$^{c\text{-}src}$ abundance and *c-src* RNA expression in human neuroblastoma variants. Oncogene 1989; 4: 421-427.
58. Bjelfman C, Meyerson G, Cartwright CA et al. Early activation of endogenous pp60^{src} kinase activity during neuronal differentiation of cultured human neuroblastoma cells. Mol Cell Biol 1990; 10: 361-370.
59. Meyerson G, Pfenninger KH, Pahlman S. A complex consisting of pp60$^{c\text{-}src}$/pp60$^{c\text{-}srcN}$ and a 38kDa proteinis highly enriched in growth cones from differentiated SH-SY5Y neuroblastoma cells. J Cell Sci 1992; 103: 233-243.
60. Maness PF, Aubry M, Shores CG et al. *c-src* gene product in developing rat brain is enriched in nerve growth cone membranes. Proc Natl Acad Sci USA 1988; 85: 5001-5005.
61. Bolen JB, Veillette A, Schwatrz AM et al. Activation of pp60$^{c\text{-}src}$ protein kinase activity in human colon carcinoma. Proc Natl Acad Sci USA 1987; 84: 2251-2255.
62. Bolen JB, Veillette A, Schwartz AM et al. Analysis of pp60$^{c\text{-}src}$ in human colon carcinoma and normal human colon mucosal cells. Oncogene Res 1987; 1:149-168.
63. DeSeau V, Rosen N, Bolen JB. Analysis of pp60$^{c\text{-}src}$ tyrosine kinase activity and phosphotyrosyl phosphatase activity in human colon carcinoma and normal human colon mucosal cells. J Biol Biochem 1987; 35: 113-128.
64. Takekura N, Yasui W, Yoshida K et al. pp60$^{c\text{-}src}$ protein kinase activity in human gastric carcinomas. Int J cancer 1990; 45: 847-851.
65. Cartwright CA, Meisler AI, Eckhart W.Activation of the pp60$^{c\text{-}src}$ protein kinase is an early event in colonic carcinogenesis. Proc Natl Acad Sci USA 1990; 87: 558-562.
66. Cartwright CA, Kamps MP, Meisler AI et al. pp60$^{c\text{-}src}$ activation in human colon carcinoma. J Clin Invest 1989; 83: 2025-2033.
67. Marian B, Winawer S, Friedman E. Tyrosine phosphorylation of a Mr 63,000 protein induced by an endogenous mitogen in human colon carcinoma cells, but not in normal colonocytes. Cancer Res 1989; 49: 4231-4236.
68. Termuhlen PM, Curley SA, Talamonti MS et al. Site-specific differences in pp60$^{c\text{-}src}$ activity in human colorectal metastases. J Surg Res 1993; 54: 293-298.
69. Weber TK, Steele G, Summerhayes IC. Differential pp60$^{c\text{-}src}$ activity in well and poorly differentiated human colon carcinomas and cell lines. J Clin Invest 1992; 90: 815-821.
70. Schwartz B, Avivi C, Lamprecht S. Isolation and characterization of normal and neoplastic colonic epithelial cell populations. Gastroenterology 1991; 100: 692-702.
71. Schwartz B, Fraser BM, Levy J et al. Differential distribution of protein kinases along the crypt-to-lumen regions of rat colonic epithelium. Gut 1988; 29: 1213-1221.
72. Burgess DR, Jiang W, Mamjiwalla S et al. Intestinal crypt stem cells possess high levels of cytoskeletal-associated phosphotyrosine-containing proteins and tyrosine kinase activity relative to differentiated enterocytes. J Cell Biol 1989; 109: 2139-2144.
73. D'Emilia JC, Mathey-Prevot B, Jaros K. Preneoplastic lesions induced by *myc* and *src*

oncogenes in a heterotypic rat colon. Oncogene 1991; 6: 303-309.
74. Pories S, Jaros K, Steele G et al. Onogene-mediated transformation of fetal rat colon in vitro. Oncogene 1992; 7: 885-893.
75. Schwartz B, Cagnano E, Braun S et al. Characterization of tyrosine protein kinase associated with subcellular components of human colonic epithelium. Anticancer Res 1990; 10: 1747-1754.
76. Garcia R, Parikh NU, Saya H et al. Effect of herbimycin A on growth and pp60$^{c\text{-}src}$ activity in human colon tumor cell lines. Oncogene 1991; 6: 1983-1989.
77. Mazurenko NN, Kogan EA, Zborovskaya IB et al. Expression of pp60$^{c\text{-}src}$ in human small cell and non-small cell lung carcinoma. Eur J Cancer 1992; 28: 372-377.
78. Jankowski J, Coghill G, Hopwood D et al. Oncogenes and onco-suppressor gene in adenocarcinoma of the oesophagus. Gut 1992; 33: 1033-1038.
79. Loganzo F, Dosik JS, Zhao Y et al. Elevated expression of protein tyrosine kinase *c-yes*, but not *c-src* in human malignant melanoma. Oncogene 1993; 8: 2637-2644.
80. Mole SE, Mulligan LM, Healey CS et al. Localization of the gene for multiple endocrine neoplasia type 2A to a 480 kb region in chromosome band 10q11.2. Hum Mol Genet 1993; 2: 247-252.
81. Lairemore TC, Dou S, Chi et al. A 1.5 megabase yeast artificial chromosome contig from human chromosome 10q11.2 connecting three genetic loci (RET, D10s94, and D10S102) closely linked to the MEN2A locus. Proc Natl Acad Sci USA 1993; 90: 492-496.
82. McDonald H, Smailus D, Jenkins H et al. Identification and characterization of a gene at D10S94 in the MENSA2A region. Genomics 1992; 13: 344-348.
83. Pierotti MA, Santoro M Herrman MA et al. Characterization of an inversion on the long arm of chromosome 10 juxtaposing D10S170 and RET and creating the oncogenic sequence RET/PTC. Proc Natl Acad Sci USA 1992; 89: 1616-1620.
84. Santoro M, Sabino N, Ishazaki Y et al. Involvement of the RET oncogene in human tumors: specificity of RET activation to thyroid tumors. Brit J Cancer 1993; 68: 460-464.
85. Donis-Keller H, Dou S, Chi D et al. Mutations in the RET proto-oncogene are associated with MEN2A and FMTC. Hum Mol Genet 1993; 2: 851-856.
86. Mulligan LM, Kwok JB, Healey CS et al. Germ-line mutations of the RET proto-oncogene in multiple endocrine neoplasia type 2A. Nature (London) 1993; 363: 458-460.
87. Miyazaki K, Asai N, Iwashita T et al. Tyrosine kinase activity of the RET proto-oncogene products in vitro. Biochem Biophys Res Comm 1993; 193: 565-570.
88. Bogarzone I, Monzini N, Borello MG et al. Molecular chracterization of a thyroid tumor-specific transforming sequence formed by the fusion of the RET tyrosine kinase and the regulatory subunit RI alpha of cyclic AMP-dependent protein kinase A. Mol Cell Biol 1993; 13: 358-366.
89. Lanzi C, Borello MG, Bongarzone I et al. Identification of the product of two oncogenic rearranged forms of the RET proto-oncogene in papillary thyroid carcinomas. Oncogene 1992; 7: 2189-2194.

CHAPTER 7

PERSPECTIVE

The role of signal transduction mechanisms in the etiology of cancer is only just beginning to be understood.[1] In the preceding chapters I have described how studies on Rous sarcoma virus have resulted in crucial insights into the role of tyrosine-specific phosphorylation in the disruption of both the regulated proliferation of cells and the ordered arrangement of the cytoskeleton. Over the past few years modern molecular biology has now allowed the identification of growth factors and receptors which also converge with similar regulatory pathways within the cell. The identification of those specific pathways leading to cell proliferation may well lead to the identification of specific intracellular targets for the development of drugs for anti-cancer therapy. For example, if there is a specific activation of *c-src* in breast cancer, as has been postulated,[2] then interfering with the interaction of *c-src* with other signaling molecules, such as the PDGF receptor or *shc*, might result in a specific inhibition of these tumor cells with no effect on other cell types, which appear not to need *c-src* for normal cellular function. Similarly, in tumors where there is an aberrant expression of growth factors, rather than blocking the growth factor-receptor interaction, it might be possible to block the interaction of the cytoplasmic domain of the activated growth factor receptor with other downstream signaling molecules to inhibit aberrant cell proliferation. A number of inhibitors of tyrosine kinases have now been developed, and a series of compounds based on phosphotyrosine have been synthesized which have differential specificities for different tyrosine kinases.[3] These tyrophostins have been used to inhibit growth factor-induced proliferation in vitro and may have efficacy against neoplasia in vivo. At present there have been no reports on the use of tyrosine kinase inhibitors in clinical trials, although it can only be a matter of time before such investigations are made public.

One area which has been largely ignored until relatively recently is that of phosphotyrosine phosphatases (PTPases). It is now recognized that these enzymes have crucial roles in regulating the extent of tyrosine phosphorylation with the cell.[4] Over 30 PTPases have been characterized and in addition to a general role for suppressing the level of intracellular tyrosine phosphorylation, many PTPases may have specific substrates and have differential tissue distribution; therefore they can be involved in specific signaling pathways. Furthermore a subfamily of PTPases has SH2 domains, and these enzymes are themselves regulated by phosphorylation/dephosphorylation events.[5] The importance of PTPases in cell regulation is underscored by studies on the T-cell specific PTPase CD45. CD45-deficient T cells are unable to transduce a signal from the T-cell receptor, but signaling can be reconstituted by transfection of CD45.[6] Different PTPases are expressed by different tissues, for example PTP1C is restricted to hemopoietic cells whereas others such as *syp* and PTP1D associate with activated growth factor receptors.[7] An essential role for PTPases in oncogenesis is indicated by a report that overexpression of PTPα (which can dephosphorylate Tyr 527 of *c-src*) results in activation of *c-src* associated tyrosine kinase activity and also neoplastic transformation.[8] Therefore interfering with PTPase function may be a viable way of inhibiting tyrosine kinase-mediated oncogenesis. One possibility for cancer therapy might be to upregulate or downregulate the expression or activity of specific PTPases involved in particular signaling pathways.

Transcription factors are at present the subject of intensive study. The activation of cells either by growth factors or by transformation by *src* can result in phosphorylation of at least two transcription factors, probably due to an activation of mitogen activated protein (MAP) kinases.[9] An important molecule in one of these signaling pathways is p91, a cytoplasmic transcriptional factor whose DNA-binding capability is regulated by tyrosine phosphorylation.[10] Recently it has been shown that in addition to activation by EGF p91 is also activated when cells are stimulated by interferons α or γ, and the kinases responsible for this activation have been identified as members of the tyrosine kinase JAK family.[11,12] Thus tyrosine kinases can be involved at multiple sites in signaling pathways, from membrane-bound receptors to nuclear events including transcriptional regulation. Moreover, these studies demonstrate that the mechanisms of signaling by a number of unrelated factors converge into common pathways within cells. Therefore the kinases responsible for transcription factor regulation will constitute other important targets for anticancer therapy. These same transcription factors are also central to the control of differentiation and so over the next few years it is likely that the mechanism of induction of a number of differentiation-specific genes in response to tyrosine kinases will be elucidated.

Many cytoskeletal proteins have now been identified and their role in normal and neoplastic cell function is being intensely investigated. Changes in the expression, function and intermolecular interactions of cytoskeletal proteins during transformation are being studied at the molecular level. In particular, the integration of cytoskeletal changes with altered tyrosine phosphorylation, and the role of small GTP-binding proteins such as *rho* and *rac* in regulating cytoskeletal changes should lead to a greater understanding of the structural changes which accompany transformation. Whilst the presence of small GTP-binding proteins are essential for cytoskeletal reorganization their exact mode of action is currently unknown. It is likely that the integration of cytoskeletal responses with signaling and second messengers will continue to be the focus of intense investigation. Current examples of this field of research are the studies showing that α-actinin functions as an actin-binding protein only when bound to PIP_2[13] and the activation of PLCγ-mediated hydrolysis of profilin-bound PIP_2 by tyrosine phosphorylation as discussed in chapter 4. PI3'kinase-mediated regulation of the cytoskeleton is another example of signaling molecules whose activation results in an alteration of cellular ultrastructure. There have been several reports describing the necessary activity of PI3 kinase for the induction of PDGF-receptor induced-cytoskeletal changes in fibroblasts or cytokine-mediated changes in leucocytes.[14,15] Although this mechanism has also yet to be fully elucidated there is a good correlation between the increased levels of PIP_3 (the physiological product of PI3 kinase activity) and changes in actin polymerization in neutrophils, suggesting that PIP_3 may regulate actin function.

It seems likely that many of these disparate signals and responses will be the result of only a few signaling pathways into which a variety of receptors and other molecules will feed. It is to be hoped that a greater understanding of these bifurcating signaling pathways and responses will be useful in the eventual understanding of human cancers and will lead to better and more specific therapies for neoplastic diseases.

References

1. Brugge JS. New intracellular targets for therapeutic drug design. Science 1993; 260: 918-919.
2. Ottenhoff-Kalff AE, Rijksen G, van Beurden EA et al. Characterization of protein tyrosine kinases from human breast cancer: involvement of the *c-src* oncogene product. Cancer Res 1992; 52: 4773-4778.
3. Levitsky AL. Tyrphostins: tyrosine kinase blockers as novel antiproliferative agents and dissectors of signal transduction. FASEB J 1992; 6: 3275-3282.
4. Walton KM, Dixon JE. Protein tyrosine phosphatases. Ann Rev Biochem 1999; 62: 101-120.
5. Vogel W, Lammers R, Huang J et al. Activation of a phosphotyrosine phosphatase by tyrosine phosphorylation. Science 1993; 259: 1611-1614.
6. Koretsky GA. Role of the CD45 tyrosine phosphatase in signal transduction in the immune system. FASEB J 1993; 7: 420-426.
7. Feng G, Hui C, Pawson T. SH2-containing phosphotyrosine phosphatase as a target of protein tyrosine kinases. Science 1993; 259: 1607-1611.
8. Zheng XM, Wang Y, Pallen CJ. Cell transformation and activation of $pp60^{c\text{-}src}$ by overexpression of a protein tyrosine phosphatase. Nature (London) 1992; 359: 336-339.
9. Marais R, Wynne J, Treisman R. The SRF accessory protein Elk1 contains a growth factor-regulated transcriptional factor activation domain. Cell 1993; 73: 395-406.
10. Ruff-Jamieson S, Chen K, Cohen S. Induction by EGF and interferon-γ of tyrosine phosphorylated DNA binding proteins in mouse liver nuclei. Science 1993; 261: 1733-1736.
11. Shesai K, Ziemicki A, Wilks AF. Polypeptide signaling to the nucleus through tyrosine phosphorylation of JAK and Stat proteins. Nature (London) 1993; 366: 580-583.
12. Silvennoinen O, Ihle JN, Schlessinger J. Interferon-induced nuclear signaling by JAK protein tyrosine kinases. Nature (London) 1993; 366: 583-585.
13. Fukami K, Furuhashi K, Inagaki M et al. Requirement of phosphatidylinositol 4,5-bisphosphate for α-actinin function. Nature (London) 1993; 359: 150-152.
14. Wymann M, Arcaro A. Platelet-derived growth factor-induced phosphatidylinositol 3-kinase activation mediates actin rearrangements in fibroblasts. Biochem J 1994; 298: 517-520.
15. Arcaro A, Wymann MP. Wortmannin is a potent phosphatidylinositol 3-kinase inhibitor: the role of phosphatidylinositol 3,4,5-trisphosphate in neutrophil responses. Biochem J 1993; 296: 297-301.

INDEX

Page numbers in italics denote figures (f) and tables (t).

U

V

W

Y

Z